Managing Natural Science Collections

Managing Natural Science Collections demonstrates the need for consistency and evidence-based decision making in the management of natural science collections, which are becoming increasingly valuable when it comes to addressing societal challenges.

Drawing upon the experience of four experts who have managed some of the largest and most diverse collections in the world, the book aims to assist in the making of strategic and operational decisions regarding care, development, access and resource management. Encouraging the reader to consider how collection strategies can be aligned with the mission of their institution and contribute to its vision, the authors also examine ways to deliver a consistent approach that will secure the present and future availability and relevance of collections. Principles of good practice and resource optimisation in an ethical and legal context are provided throughout the book, as well as case studies, sample documents and templates, all of which will be useful for discussion and teaching.

Managing Natural Science Collections encourages each reader to consider the different options available to them. As such, it should be essential reading for museum practitioners and other professionals around the world who are involved with any strategic aspect of managing natural science collections. Students of museum studies will also find much to interest them within the pages of this book.

Robert Huxley was Head of Botanical Collections at the Natural History Museum (NHM) in London for 21 years, where he led the team managing the NHM's 5.2 million plant and fungal specimens. He is now a Scientific Associate at the NHM, an Honorary Curator at World Museum, Liverpool and a Guest Researcher at the Museum für Naturkunde, Berlin (MfN).

Christiane Quaisser has worked on strategic management of natural science collections for more than ten years. In 2013, she became Head of Collections at the Museum für Naturkunde, Berlin (MfN), leading the change into the new Science Programme, Collection Development and Biodiversity Discovery.

Carol R Butler has led the Smithsonian's National Museum of Natural History collections work for more than ten years, rising to the position of Assistant Director for Collections in 2012. Her current interests are in collections assessment, developing means of describing collections as resources, encouraging best practices in tracking the uses of collections and in developing communities for broad collaborations.

René WRJ Dekker has 35 years of experience in research and management of natural history collections, as Scientist, Curator and Director of Collections. He has authored over 80 papers and a book on megapode birds (*The Megapodes*, 1995). He has now returned to research as Senior Scientist in Ornithology at the Naturalis Biodiversity Center.

Managing Natural Science Collections

A Guide to Strategy, Planning and Resourcing

Robert Huxley, Christiane Quaisser, Carol R Butler and René WRJ Dekker

Routledge
Taylor & Francis Group
LONDON AND NEW YORK

NATURAL HISTORY MUSEUM

Naturalis Biodiversity Center

für Natur
MUSEUM FÜR NATURKUNDE
BERLIN

Smithsonian
National Museum of Natural History

First published 2021
by Routledge
2 Park Square, Milton Park, Abingdon, Oxon OX14 4RN

and by Routledge
52 Vanderbilt Avenue, New York, NY 10017

Routledge is an imprint of the Taylor & Francis Group, an informa business

British Library Cataloguing-in-Publication Data
A catalogue record for this book is available from the British Library

Library of Congress Cataloging-in-Publication Data
A catalog record has been requested for this book

ISBN: 978-1-138-38681-5 (hbk)
ISBN: 978-1-138-38683-9 (pbk)
ISBN: 978-0-429-42659-9 (ebk)

Typeset in Sabon
by Newgen Publishing UK

Contents

Figures

Tables

About the authors

Robert (Rob) Huxley was Head of Botanical Collections at the Natural History Museum (NHM) in London for 21 years where he led the team managing the NHM's 5.2 million plant and fungal specimens. Educated at Liverpool and London Universities he gained his PhD in taxonomy and ecology of marine algae and attained post-graduate qualifications in Continuing Education at Greenwich University and, in recent years, in Coaching and Mentoring Practice at Oxford Brookes University. Rob initially worked in marine science research before venturing into the museum sector, firstly managing a museum and heritage project in North Wales before joining the NHM in 1987, starting in exhibitions development then moving to collections management from 1990. Rob led and participated in many national and international projects carrying out reviews and assessments, building networks, developing new methods and providing training in collections management in several European countries, in the United States and in Kenya. Rob coordinated the networking activities of the EU-funded SYNTHESYS project (2003–2015) and led the development of the EUColComp competency framework. Along with René Dekker and Christiane Quaisser, Rob was a member of the Collections Policy Board of the Consortium of European Taxonomic Facilities (CETAF). Rob has been a member of the board of CETAF and the Natural Science Collections Alliance (NSCA). From 2002–2004 Rob was President of the Society for the Preservation of Natural History Collections and now sits on the editorial boards of the Journal of Natural Science Collections and the Journal of the History of Collections. He has a strong interest in the history of natural history leading to publications such as *The Great Naturalists*. He is now a Scientific Associate at the NHM, an Honorary Curator at World Museum Liverpool and a Guest Researcher at the Museum für Naturkunde, Berlin.

Christiane Quaisser has worked on strategic management of natural science collections for more than ten years. In 2013, she became Head of Collections at the Museum für Naturkunde, Berlin (MfN) leading the change into the new Science Programme, Collection Development and

Biodiversity Discovery. Christiane is a biologist by profession and has a background in ornithology. After her PhD on the impact of agricultural land-use on the breeding success of birds, she started her museum career as a trainee in the bird collection of the natural history museum in Dresden (SE Germany) in 1999. Scientific projects on nomenclatural and taxonomic problems with types brought her to major bird collections in Europe for research studies, e.g. to Leiden, Paris, Vienna and London. The deep insights in these collections inspired her to move on to collections management, first as a project coordinator at Naturalis in Leiden, and then from 2009 onwards at the MfN in Berlin. From the beginning, Christiane's work has focussed on strategies in management and development of natural science collections to improve their standard, accessibility and relevance. Together with René Dekker, Rob Huxley and Michel Guiraud, she initiated a European working group of Directors of Collections and was involved in European projects, e.g. SYNTHESYS, EDIT, EUColComp, laying the ground for many aspects of the book. Together with Carol Butler, Christiane has been engaged with a collection working group of 12 natural history museums and further development of collection assessments. Christiane was member of the evaluation panel of natural history museums in Brussels and Florence and of the Belgian heritage science programme BRAIN-be 2.0. Current activities are dedicated to a comprehensive collection development plan for the MfN in the context of societal changes. Current research interests are focussing on conservation science, e.g. deteriorations processes in mammal skins. Both are underpinning the importance of interdisciplinary collaboration as a guiding principle and key success factor.

Carol R Butler has led the Smithsonian's National Museum of Natural History (NMNH) collections work for more than ten years, rising to the position of Assistant Director for Collections in 2012. She began her museum career during her graduate school programme in anthropology at Wake Forest University, United States, where she developed and presented education programmes and exhibits, wrote annual reports and inventoried the collection and planned its move into a new facility. She did fieldwork at the Neolithic archaeological site in 'Ain Ghazal in Amman, Jordan. Early work at NMNH included inventories of anthropological collections and detailed space planning for housing them at the Museum Support Center. She became acquainted with this volume's co-authors through engagement in the SYNTHESYS assessment pilots and meetings of the Society for the Preservation of Natural History Collections. Carol's work now focusses on leading the direction and management of NMNH's anthropological, biological, geological and archival collections. She leads the central teams that provide Biorepository, Conservation and Registration services along with technical teams that conduct collections improvement projects. Drawing upon her experience in advising employees, supervisors and management on sensitive personnel problems and staff organisation, Carol also contributes to the Smithsonian's leadership development programmes by serving as a mentor and rotational supervisor. She was the lead designer and faculty member for the Smithsonian's Latin American and Caribbean Collections Management Training Program; additionally, she has provided extended training to museum professionals from Iraq, Oman, Sri Lanka, India and Pakistan. Recent work includes foundational action that created the Global Genome Biodiversity Network (GGBN). She serves on GGBN's Executive Committee and is the past Chair of its Policies Taskforce. Carol has been engaged with a collections working group of 12 natural history museums that have resulted in a high level Collections Dashboard. Her current interests are in collections assessment, developing means of describing collections as resources, encouraging best practices in tracking the uses of collections and in developing communities for broad collaborations.

René WRJ Dekker was Director of Collections at Naturalis biodiversity Center, Leiden, The Netherlands for 11 years. He was educated in biology at the Free University of Amsterdam. While researching for his PhD at the Zoological Museum of the University of Amsterdam he lived in Sulawesi, Indonesia, where he studied megapode birds. In 1991 he became Curator of Birds at the National Museum of Natural History (now Naturalis Biodiversity Center). René has 35 years of experience in research and management of natural history collections, both as Scientist, Curator and Director of Collections (DoC). As DoC he was in charge of and responsible for 60+ staff and 42 million objects and had overall responsibility for management, policies and finances as a member of the management team, with a special focus on collections. During his years as DoC, Naturalis merged with the Zoological Museum of Amsterdam and the National Herbarium. This included the merger of staff and the transfer and integration of collections. René supervised a large-scale digitisation project for five years, with a budget of 13 M Euro which enabled 75 additional staff to digitise 7 million objects. René is former chairman of the EU Director of Collections/Collection Policy Board of major European natural history museums. He has authored over 80 papers and a book on megapode birds published by Oxford University Press. He has now returned to research as Senior Scientist in ornithology at Naturalis Biodiversity Center.

Foreword

The landscape of collection management in natural history museums has significantly changed during the past 20 years. From the common perception of being merely research assistance, it has evolved its own mission with a family of job descriptions supported by many competencies. Today, managing natural science collections is something that echoes the big numbers of archaeology, the historical traits of archives, the material diversity of cultural objects and the information contained in large libraries. The users of natural science collections have also expanded from taxonomists describing nature's diversity to scientists working on socio-economic issues such as global change, health, food security or migrations. Natural science collections are heritage that needs to be preserved and passed to future generations, but they are also tools and material for supporting current research. They are needed to disseminate science to the public and must embrace the digital age by opening up their inventories and image banks. How does one mitigate between these contradicting interests? Collections reflect more than three centuries of the World's descriptions but current research produces more and new types of collections that require resources and space. How do you manage, if not prioritise, between past, present and future?

Because of these evolutions and the questions they raise it was necessary to provide collection managers with some principles and paradigms that would help them to place their work in the context of the current role of natural history museums. The authors of this book are the best placed to fulfil these needs. For many years they have held senior positions in collection management of major institutions and implemented the strategies required to meet 21st century challenges. During those years, I shared with them the building of a professional network, supported not only by our institutions but also by European and international initiatives. CETAF, the Consortium of European Taxonomic Facilities, was the bedrock of our association, as the convergence point of the European infrastructure programmes SYNTHESYS and EDIT (the European Distributed Institute of Taxonomy). Through these programmes, the necessity emerged for a board of directors of collections to elaborate a common vision across our institutions. That board existed as the Collections Policy Board in which Christiane Quaisser took a very active part and together with Rob Huxley and René Dekker led the group to develop guidelines and write position papers. The network activities of the SYNTHESYS programme have been extremely fruitful for collection management. A main achievement in this regard was the assessment of European collections in which Robert Huxley, Christiane Quaisser and Carol Butler were involved, and which led Robert to design a collections self-assessment tool. A second EU significant project led by Robert Huxley with Christiane Quaisser and René Dekker developed EUColComp, a competency framework aimed at supporting human resource management and staff training, and understanding the job description in collection management over Europe in order to facilitate staff mobility across institutions. Carol Butler and Robert Huxley applied this methodology at the Smithsonian, exemplifying the generic value of the tools developed in this network, and reciprocally, Carol Butler was instrumental in the development of the collections assessment schema known informally as "move the dots" which has been applied in US and European museums. It was developed to answer basic questions about the appropriateness of the collection relative to the museum's mission, and the readiness of the collection for use.

The reader will appreciate the comprehensive nature of this book and how it addresses all the issues that a head of collections or a senior collection manager has faced or will face during his or her career. It will make him or her a "doer" as much as a "thinker" and the long list of questions to ask oneself which is provided in the last chapter will certainly appeal to anyone dealing with the challenges and opportunities that natural history collections present.

Prof. Michel Guiraud
Assistant Director General for Collections
Muséum national d'Histoire naturelle
Paris, France

Acknowledgements

We are grateful to all our colleagues from Europe, the United States and worldwide who have engaged with us in many working groups and projects, shared their ideas and helped us along the way to develop and implement common principles, methods and standards and to raise awareness and develop a professional approach to strategic management of natural science collections. Without them this book would not have been possible! It would be impossible to single them all out but special thanks must go to the following:

Michel Guiraud, Assistant Director General for Collections at the Muséum national d'Histoire naturelle, Paris, France. Without our dear friend Michel, this book, and our lives, would be lacking in many aspects gained through those unforgettable, inspiring meetings facilitated by good wine and good food.

Johannes Vogel, Director General Museum für Naturkunde, Berlin. The idea for this book project dates back to a time when Johannes was Keeper of Botany at the Natural History Museum, London. From this very first moment until today, through many changes over time, he has always supported and enabled this project in many different ways.

Petra Ebber, Assistant, Research Programme Collection Development and Biodiversity Discovery, Museum für Naturkunde, Berlin. The most important people are often the ones that are hardly seen. Petra helped us with the logistics in Berlin, finding hotels, organising reimbursements and taking good care of us during so many meetings at the Museum für Naturkunde.

Colleagues at the Natural History Museum, London for support, encouragement and patience! Particularly Trudy Brannan, Editorial Manager; Vincent Smith, Research Leader, Informatics, Diversity, and Informatics Division for his help with Chapter 7; and David Williams, Researcher, Diatoms, Algae, Fungi and Plants Division. All those who contributed to the development of the competency framework on which EUColComp was based.

Thanks also go to Caroline Pepermans, Steven van der Mije, Luc Willemse, Christel Schollaardt, Aimee van Schie and staff of the Collections Department of Naturalis Biodiversity Center for their stimulating discussions, hard work and involvement over the years and like other friends and colleagues who, over three decades, made life at Naturalis a pleasant experience. Marc Gofferjé, project manager during the initial phase of the digitisation project at Naturalis, is acknowledged for his contribution to Chapter 7 of this book.

Director and Deputy Director of World Museum Liverpool, Steve Judd and Anne Fahy and their teams for providing us with facilities and support to work on the book away from the distractions of our everyday jobs. And for other colleagues at National Museums Liverpool and Victoria Gallery and Museum, Liverpool for testing out our list of strategic questions: Wendy Atkinson, Sophie Banks, Donna Young, Tony Hunter, Tony Parker, Geraldine Reid, Leonie Sedman, Wendy Simkiss and Maureen Smith.

Kirk Johnson (Director) and Ian Owens of the Smithsonian National Museum of Natural History for their support and encouragement, and many more colleagues, especially Kelly Tomajko, Tim White, Linda Ford, Bob Corrigan, Jonathan Coddington, Bill Tompkins, Clare Valentine, Cathy Hawks, Deb Hull-Walski, Sarah Loudin, Kathy Hollis and Chris Huddleston for the many helpful conversations along the way.

And finally, but not least, the European Commission for supporting many of the projects that we have worked on together to develop the tools and guidelines included in this book.

1 Introduction

You have just taken up a job in a museum as project manager leading a team with responsibility for the digitisation of a collection of some 300,000 natural science objects preserved in a variety of ways. Your director is expecting results on schedule and within budget, making the collection accessible to a wide range of users. Or perhaps your background lies elsewhere in the museum sector and this natural science collection is part of a broader collection for which you have responsibility. Or, maybe you have spent some years as a researcher using the collections and now in your new management position find yourself responsible for their preservation and access to users. Whichever of these or many more scenarios fit your personal circumstances you will be confronted with a baffling array of questions.

Where do you start? There may be help from colleagues or there may not and perhaps you are the only natural scientist in the institution. You know that you need to take a strategic approach and not be drawn into detail – that can come later or be the job of your team if you have one. These are the positions that we, the authors, were once placed in. We all agree that we would have loved it if there had been a book that would have helped us to ask the right questions and encouraged dialogue with colleagues to make real changes. We hope that this is that book and that we can share our experiences and some of the methodologies and tools that we have developed and used during our careers. Many of the challenges we experienced were related to the nature of natural science collections themselves, their vast size and the way in which they are used. Other challenges were with the individuals with a wide range of skills, knowledge and world views that we were now to manage. Museum workers from other disciplines are often amazed if not shocked by the tens of thousands of specimens that are traveling around the world in the mail as loans from museums to researchers and back. Natural science collections do share many characteristics with other museum collections and are a popular element for museum display and education as are art and humanities objects. However, those on display are a small fraction of the vast resource held in storerooms around the world. In this book we will focus on natural science, but we hope that many of the questions we pose and answer will be applicable elsewhere.

What is special about natural science collections?

While natural science collections share many characteristics with other museum collections, they have a particular combination of characteristics that present unique challenges to their preservation and opportunities for their use. They tend to be high in numbers (for example the largest collection is that of the Smithsonian National Natural History Museum with 146 million specimens); they are composed of natural material rather than human manufactured objects; they are highly diverse in both material and size; most were created as a research infrastructure (curiosity and research driven) supporting and linked to active research; they are mostly made up of unique individuals so for example, like human beings, none of the millions of beetles in a particular collection is a copy of another one; their labels hold data that provides a record of the existence of an organism in space and time. Some have special status as type specimens, the anchor point for scientific descriptions and names.

The changing landscape

The acquisition, care and management of natural science collections have always been challenging. From the activity of collecting specimens in extreme environments or from distant and unfamiliar countries, to satisfying the exacting preservation requirements of specimens that would otherwise naturally deteriorate, collections management has always required specific knowledge and skills.

The landscape of increasing demands for virtual and physical access, technical opportunities and challenges, as well as constrained resources, drives collections professionals to continually adjust and tune their approaches.

Collections of dried plants, pinned insects, fluid-preserved invertebrates, stuffed vertebrates, fossils, minerals, rocks and maybe even ice cores, described as natural science collections, are housed and mostly cared for in a variety of institutions around the world from the large purpose-built natural history museums, through herbaria of botanic gardens, to university collections and local, often multidisciplinary museums. These natural science collections stimulate curiosity and wonder, leading to scholarly enquiry and discovery as well as learning and enjoyment through exhibition and education programmes for all ages.

In the last 25 years, natural science collections have increasingly been viewed collectively as a vast distributed infrastructure for research and discovery, contributing to our understanding of the diversity of life and providing a source of baseline data for research from human and animal health to environmental and climate change. This change in perception mainly reflects a shift of research focus to more applied science with societal relevance. New scientific questions arise, driven by the transformation of physical entities to digital surrogates from multiple source collections across the world which can be integrated, repurposed and presented through joint web portals.

Challenges of natural science collections

Natural science collections number in the tens of thousands to millions in institutions around the world. The availability of funds to care for, house and make them accessible to the expanding user base has diminished significantly, putting many, especially smaller collections, at risk of loss or dispersal. Their scale and changing usage present new challenges to those responsible for their care, development and accessibility. There is a greater need for collaboration, common approaches and joining forces, and consequently there is a need for greater understanding of strategic management and an understanding of their changing position in science and society. At the same time, individual specialist curators with in-depth knowledge of the natural world have become an increasing rarity. Although the same trends can be seen in arts and humanities collections, natural science collections have many specialist qualities and needs. This presents different challenges to managers of these collections, whether they are familiar with natural sciences or are from other academic backgrounds.

The scope of this book

In this book we aim to share our experiences developing as managers in four large natural science museums and also some of the products of the various international collaborative projects that we have worked on over the last 20 years. We have had close involvement with specialist societies such as the Society for the Preservation of Natural History Collections (SPNHC), and through collaborative projects we have led and participated in surveys and reviews of museums large and small in many countries. This experience has led us to believe that the principles described in this book can be applied at all levels, from the largest nationals to small local museums with a single staff member and a team of volunteers, or to specific collections projects such as large scale digitisation or a major rehousing. We hope that it will assist you in making strategic decisions regarding care, development and access and indirectly to encourage a consistent approach to managing natural science collections to ensure their present and future availability and relevance. Regardless of your personal position and the governance and mission of your institution, it aims to help you to gain confidence, build experience, and, importantly, to know what questions you need to ask when confronted with drawers of insects, cabinets of plants and boxes of fossils of varying states of preservation.

Principles of good practice and resource optimisation in an ethical and legal context are considered throughout the book, using case studies, sample documents and templates, all of which can be used for discussion and teaching, and links to references. We aim to pose questions and give answers from our perspective but at the end of the day, how you will answer them will depend on your own circumstances, vision and objectives.

Although we will direct you to specific references and general reading, much of what we share with you below is based on our many years of experience and knowledge as scientists and museum professionals. Many terms, acronyms and concepts are included in the Glossary.

We do not aim to cover operational aspects of managing natural science collections as there is a wealth of literature available and excellent resources developed through specialist societies. Topics that apply across the museum sector, such as documentation, are expertly dealt with by the Spectrum standard for example (https://collectionstrust.org.uk/spectrum/). Preventative conservation (which ultimately underpins all the processes we describe) has in the last 20 years received the attention it deserves in the natural sciences and, again, there is an extensive literature available. We would particularly draw your attention to the excellent SPNHC publication *Preventive Conservation: Collection Storage* (Elkin and Norris, 2019). This substantial and comprehensive volume was published on the cusp of our submitting this volume for publication so you will not see any specific references but resources such as this are part of our continuing learning as well as yours.

In our book we hope to address the need for consistency and quality decision making in the management of these large, highly variable and increasingly valuable collections.

Staff who formerly had limited areas of responsibility now need to make broad, strategic decisions but do not necessarily have the knowledge or training at the required level. A new type of collections manager is evolving with much wider responsibility and competence. We aim to raise awareness of new managers and established specialists alike to some of the tools, methods and approaches now available. Our ultimate aim is to prompt discussion rather than be over-reliant on existing facts. In the following paragraphs we will give you a summary of the areas that we cover in the next eight chapters.

Why are we writing this book?

We came to collections management from research or other academic backgrounds, mainly as users of collections ourselves. We had been practicing anthropologists and taxonomists in some of the world's largest museums, working with the collections as the raw material to understand, classify and name our small pieces of the natural science jigsaw puzzle. We had all worked as collections managers or registrars and when we took management roles we all faced significant challenges. We joined at a time when there was in many museums a move to "put the management in collections management", treating the management of our natural science collections through a business management lens. This was also a time when the funding for natural science and many other collections had been dwindling for many years and where our collections needed to justify their very existence.

We had all participated in reviews and surveys of collections and their management in a wide range of museums and herbaria mainly in Europe and the US and also our own institutions. We collectively noted that while the uses of collections by researchers were much the same, the ways in which they were managed, preserved and made accessible varied considerably within and between institutions. Accepted practice often overruled good practice and research drivers overruled collections care. How could this be redressed? We participated in international consortia and projects such as the Consortium of European Taxonomic Facilities (CETAF), the Collection Policy Board of CETAF (CPB) where we developed, among others, common guidelines for access and loans, and SYNTHESYS (www.synthesys.info/), where we were able to run courses to raise awareness of strategic thinking and new methodologies in the profession. In this book, we draw on these initiatives and the knowledge that we built up through sharing with our colleagues and, importantly, the things that we had to learn from scratch. We see this book as part of that process and something that complements these initiatives on a personal level. This is the book that we wish we had had when moving into management years ago.

What kind of collections and associated work do we address in this book?

This book primarily addresses natural science collections and includes traditionally preserved biological and geological specimens such as pinned insects, invertebrates in spirit, dried plants and mineral and fossil collections. This is not a manual on how to preserve and conserve plants and animals, how to set up a monitoring programme or how to achieve best practice in establishing museum documentation. Rather, we cover their management in museums, universities, herbaria and botanical gardens. We also include collections specifically preserved for molecular use such as frozen tissue and DNA, but only those housed and managed alongside traditional collections. However, while most of our experience, and hence the focus of the book, is on these collections, the principles extend to purpose-built biobanks of frozen tissues and living collections such as zoos, living plant collections in botanical gardens, viral, bacteria, cell and fungal cultures.

A summary of contents of this book

In **Chapter 2** we explore the background to natural science collections, their historical development, changing value and usage and lead into their modern role as a physical and virtual infrastructure for a broad user community. Understanding their value beyond cultural artefacts is key to advocating for these collections and often one of the most rewarding parts of our jobs. Once the context has been established, we invite you to consider the arguments for the expenditure of time and resources beyond the cultural significance of your collections – what is the long-term responsibility of those managing collections and how has the context and ethical basis of collections changed? We give a brief summary of how natural science collections have been accumulated over 300 or more years and how they are used. Their role has broadened in recent years from describing, classifying, identifying and deciphering the evolutionary history of the natural world to supporting a range of vital scientific research from climate change to forensics and as a source for understanding the history of science. We are increasingly expected to be advocates for our collection and that could be deemed to be a responsibility as well as a duty. We give an overview of case studies and proof of concept projects that are available to help you justify the maintenance and housing of what can be an expensive enterprise. There are great benefits to be gained by natural science collections from forming partnerships and collaborations and examples range from sharing best practice to developing common access mechanisms.

In **Chapter 3** we look at the advantage of strategic planning to achieve your goals, whether for specific areas such as conservation, more general plans for the whole institution or maybe an outline for a specific project. We emphasise the value of objective data over anecdotes and decision making based on the reality of sound data and illustrate the progression of strategy to application. What are your collections' strengths and weaknesses or importance? We look at some of the many tools available for assessing collections to aid in prioritisation and to back up your case for expenditure of time, money and resources. We encourage you to think about why an assessment is needed, what and how it is being assessed and how data are presented to different audiences to gain maximum impact. We list the key elements of a plan, give some suggestions as to how the process can be made easier and give an example of the challenge of prioritising scarce resources over equally as important collections. Examples of existing strategies and planning exercises are given for you to examine critically. We look at measuring improvements over time and take a new look at well-established tools such as performance indicators and examine what makes a good indicator.

The increasing legal and ethical considerations that affect all collections are discussed in **Chapter 4**. We discuss how governance, which is an array of laws, regulations, polices and ethics might affect your strategic planning process and how they can help you achieve your goals. Policies in particular are a key element in minimising risks to collections, staff and institutional reputation. We also consider what makes a good policy and what is appropriate for natural science collections compared to other disciplines such as art. Collections organisations need to be aware of all these and ensure that they are carrying out their business in a legal and ethical way.

In **Chapter 5** we explore the human element of our strategic approach. Collections rarely exist in isolation without any form of contact and interaction with people. They require expert input to ensure their long-term survival and maximise their accessibility, but very importantly to add and update the information associated with them. This added value is provided by individuals such as curators, collection managers, researchers, amateur specialists and volunteers. We look at some tools for identifying the competencies needed by your staff and volunteers to manage your collection or carry out specific projects such as digitisation. The way in which collection organisations are staffed and organised can affect their effectiveness to serve user needs and ensure the long-term preservation of the collections. Managers need to know if they have the array of competency at an appropriate level to achieve their strategic goals. Many tools for staff development exist. We focus on competency frameworks and some recent examples of their application in natural science collections that help managers to match competencies to the needs of the tasks to achieve their goals and develop staff. Finally, we consider how you can avoid losing vital expertise and knowledge when staff leave or move into new positions.

In **Chapter 6** we provoke thought on the strategic and ethical application of acquisition and deaccession of specimens. Natural science collections have developed in a relatively haphazard way, reflecting, amongst others, colonial expansion and fashion. In the 21st century larger collection organisations have focussed acquisition strategies to reflect the needs of research, local interest and some targeted acquisition for exhibition display. We emphasise the principle of not collecting more but collecting smarter as a key element of strategic management, for example targeted collecting and national/international development strategies. The sometimes neglected other half of this equation is what goes out? Deaccession is an essential if controversial tool in the development of collections. We are under pressures of space and resources in our institutions. So would some collections be better housed elsewhere? We look at some of the sensitive but essential questions which need to be asked and some of the tools to help the process and provide guidelines to assist organisations needing to dispose of collections. We also explore the benefits and challenges of long-term jointly held collections, managed via curation agreements, as an option for museums under resource constraints.

Chapter 7 takes a strategic view of one of the current major drivers of collections activity, the demand for internet access to data and images of collections. New technologies and industrialisation of the digitisation process present new opportunities but there remains the need to prioritise. We do not cover details of hardware or software here as this information is readily available. Rather we ask the strategic question of which collection should be digitised first, to what level of detail and, importantly, which user group will this digital collection serve?

In **Chapter 8** we invite you to question how you provide access to your collections at the strategic level. Access to collections can be by individual visits, by loans of material and by digital access via the internet. There has been an increasing shift in recent years to the latter. We encourage thinking on balancing the risks of physical damage and loss of objects and the potential gains of these means of access. We review the risks associated with loans and visits such as security and physical damage through handling. There are increasing constraints on loans and they are increasingly affected by requirements of access and benefit sharing agreements and international regulations governing "dangerous" goods. We consider common dilemmas faced by collections managers such as who gets priority and internal versus external users. Destructive sampling is included in this chapter as a particularly controversial and challenging means of access. Managers are introduced to general principles, checklists for sampling and are encouraged to think about alternatives if available. This chapter focusses on these new elements of access and the published guidelines and other material to help navigate these difficult waters.

Chapter 9 is an opportunity for you to put into practice what you have learned in the previous chapters and any resources that might help you along in your journey. We pose a series of questions in the form of a checklist that you might want to ask when planning your project or making other changes in your area of responsibility. You should be able to answer those that are relevant to your particular case either from the book or the resources identified. If you don't have the experiences related to management responsibilities yet, or if you are aiming for a new position, or if you simply want to practice, we provide you with scenarios setting the frame and starting point that might help you with developing a strategy and answering relevant questions.

Our legacy

Natural science collections are unique in many ways and they have a rich and fascinating past. To manage them in a strategic way requires a background on what they are, where they came from, how they were used and what user needs can they meet today. These collections deserve our maximum effort and professionalism to guarantee optimal availability to a large audience worldwide and for centuries to come.

Note that natural science "objects" are generally referred to as "specimens" usually depending on context and we use both terms in this book.

Reference

Elkin, L. & Norris, C.A. (eds.), 2019, *Preventive Conservation: Collection Storage*. New York: Society for the Preservation of Natural History Collections; American Institute for Conservation of Historic and Artistic Works; Smithsonian Institution; The George Washington University Museum Studies Program.

2 Natural science collections – their place in society

> Assuredly, that man who should spend his whole life in amassing together stone, timber, and other materials for building, without ever at the making any use, or raising any fabric [sic] out of them, might well be reputed very fantastic and extravagant. And a like censure would be his due, who should be perpetually heaping up of natural collections without design. Building a structure of philosophy out of them, or advancing some propositions that might turn to the benefit and advantage of the world. This is in reality the true and only proper end of collections, of observations, and natural history: and they are of no manner of use or value without it.
>
> (John Woodward, 1728)

Introduction

When naturalist and collector John Woodward wrote this, the study and collection of the natural world was developing exponentially; stimulated by the questioning approach of philosophers such as Francis Bacon and René Descartes and fuelled by the masses of new and strange objects returning to Europe from far off and newly explored lands. Unlike those who collected curiosities for their own worth, Woodward saw collecting as being towards a particular end and that without active study such collections were pointless. In this chapter we will take a brief travel through time from the first known collections up to the present day in the belief that knowledge of the history of collecting gives context to the collections that we hold and provides material for story-telling and advocacy for our natural heritage. We then invite you to consider what "value" means in the context of these collections. What is their relevance to society in the 21st century? Does Woodward's statement hold true? Are the costs of keeping and making them accessible justified? We argue that collections are an asset and that it is part of our role as collections professionals to raise their profile and be advocates for their maintenance and use.

Questions

- What are the origins of your collections: who acquired them and why?
- Why are your collections an asset and not a burden to society?
- Are there potential users who don't yet realise the value of your collections to their work? How can you engage them?
- How can you encourage political and financial support for your collections?

What are the origins of your collections: who acquired them and why?

How well do you know the origins of your collections and their global context? Although this may not at first seem to be of strategic importance, we believe that it is valuable to have a grounding in the history of collections in general and in particular regarding those collections for which you are responsible. As well as improving engagement with historical researchers, knowledge of the collection and its uses provides a basis for advocacy as we will see later in this chapter. This background information also forms part of the intellectual framework of a collection, i.e. "what we have and why we have it" and is covered in Chapter 3. Most museum professionals are aware of the history of collecting, at least within their own discipline, and for those entering the profession from a non-museum background there is a wealth of literature from the general to the specific on the fascinating individuals who created the collections that we manage today.

We can reasonably assume that pre-literate cultures would have classified plants into those that were edible, poisonous, or of medicinal value, and recognised animals as potentially dangerous or valuable. Our forebears may even have made transitory collections of examples for reference. The first formal classifications of the living world appear in the works of classical scholars such as Aristotle and there is some suggestion that Alexander the Great collected and brought back to Greece specimens of plants and animals for his old tutor to study (Freeman, 2011, p26). The collections would have served as tangible, examinable proof of the existence of creatures from beyond the known world – more reliable than any story. Whether these collections were deliberately preserved for future study is unknown. The first known herbarium was compiled in the 16th century from plants in the botanical garden of the University of Bologna for its creator, Professor Luca Ghini, to teach the characteristics and medicinal properties of plants in winter when fresh material was not available. Natural science specimens also took pride of place in the cabinets of curiosity amassed by scholars and connoisseurs where they sat alongside ethnographic objects and other curios as part of a microcosm of unfamiliar and wonderful objects – a window into new and exotic lands and peoples intended to amaze and impress contemporaries. Many of these collections eventually became more widely accessible in the first public museums. Late renaissance scholars, such as the zoologist and physician Conrad Gesner (1516–1565), saw the advantage of using specimens to communicate the concepts behind names of newly discovered species to their networks of fellow enthusiasts.

In the 17th and 18th centuries science was developing rapidly as a means to understand and interpret the natural world and the naturalists of the time were driven by curiosity mixed with a range of societal factors such as economics. The utility of collecting the natural world and the "value" of specimens led to its dramatic expansion from the mid-17th century. The "Great Collector" and successful physician Sir Hans Sloane amassed a vast collection of some 400,000 objects including plants, animals, fossils, minerals and manuscripts. Sloane and his contemporaries compared new plants and animals accrued from lands new and exotic to Europeans by collecting or by gift, exchange and purchase to be named, described and classified and this process remains much the same today. Sloane appreciated the value of these specimens to the understanding of the world and made provision in his will for the British government to purchase it at a low price for the benefit of all, thus creating the British Museum[1]. From this period geological collecting grew, playing key roles in the debates on the nature of the earth itself and not least the origin of fossils. Collections increased exponentially as the great voyages of exploration of the 18th century, like those of Alexander the Great centuries before, and brought back physical, testable evidence of the natural diversity of new lands often with an eye to its use and exploitation. It is undeniable that these new collections, which form the core of many of the world's great museums, are a reflection of a colonial past. Improved preservation techniques such as "spirits of wine" – alcohol distilled from wine, the forerunner of the methylated spirit used today – allowed soft-bodied animals and internal organs to be preserved, providing information beyond that of dried skins and skeletons. The fascinating history of these and other techniques are splendidly articulated in Stephen Asma's *Stuffed Animals and Pickled Heads* (Asma, 2001). New technologies such as microscopes gave access to the microscopic world and spurred interest in different kinds of collections. From the late 18th and through the 19th century science became increasingly professionalised as did collecting – professional naturalists began replacing gifted amateurs and well-planned and state-funded scientific expeditions opened up new environments such as the deep sea. This was the age of the public museum. Industrial towns built museums invested with civic pride, housing collections from local amateurs to the great voyages with a mandate to educate and entertain the public. The custodians of natural science collections were often gifted scientists themselves; authorities on the taxonomy of their own specialist genera and families of organisms. This tradition continued through the

20th century with many larger museums and botanic gardens becoming the world's leading research centres in biodiversity studies. In smaller and regional museums, display of specimens remained the focus and collections were often acquired as representative of local fauna and flora. Scientists, both professional and amateur, continue collecting to this day, albeit in a more sensitive and managed way than their forebears and usually linked to specific research questions (see Chapter 6).

The importance of this resource has become paramount in the 21st century, a time of change and challenges, both economic and environmental, and increasingly natural science collections are viewed as an infrastructure that can support a wide range of critical research areas. At the same time these resources are suffering from years of decline in funding and many are at risk from lack of curatorial support.

Over the last 20 years, in addition to the familiar digital revolution and the drive to mobilise digital surrogates of our collections, there has been an explosion of new techniques and new research fields. Molecular systematists are now able to recover and sequence DNA from a wide range of museum specimens including some that were considered irreparably damaged by past treatments (Hykin, Bi & McGuire, 2015). CT scanning has opened up fossil specimens to examination without damage. New techniques are delivering more data and posing new scientific questions. At the same time, we are confronted by major societal challenges such as biodiversity loss, climate change and emerging diseases and, as we will see below, natural science collections have a role to play. For the collections professional this has generated new tasks and new responsibilities, especially for the larger natural science collections, and last but not least there are new kinds of collections to manage: digital, cold/tissue/DNA collections and environmental samples.

In the next section we will look at how uses and maybe the purpose of natural science collections is changing in the 21st century and explore the arguments in support of and against them. We hope that understanding these opposing positions will help you to advocate for funding and other support for your collections.

Why are your collections an asset and not a burden to society?

"Yes, but why do you need 3000 flies of the same species?" Such questions have been asked of natural science curators many times by representatives of funding bodies, politicians and potential donors when they are confronted with drawer upon drawer of collections proudly displayed by an enthusiastic curator. The same question can be asked about rodent skins and skulls, dried plants, shrimps, ore samples and any other type of natural science specimen. This question is echoed by scientists from disciplines beyond taxonomy, although the concept of replication and statistical rigor may chime with them to some degree. The question of "why so many" is understandable, not unreasonable, and must be answered, especially as the costs of collecting and storing these collections can be significant over the decades. Plants have to be dried, insects pinned, skeletons and skins prepared, spirit collections maintained and all these protected from the agents of deterioration that can reduce them to a pile of debris of little or no value or utility. If they are not to be left as dormant, closed collections, they also require collections managers' time to send loans, supervise visitors and add new specimens to the collections. Finally, specimens can take up a lot of valuable space. Perhaps it was inevitable that questions would be asked as to whether this apparently expensive activity was worth it?

Why so many of the same thing?

To the natural scientist reading this book the answer is well-known, i.e. they are not all the same. Our collections are made up of unique individuals with important exceptions of botanical and mycological duplicates[2]. Specimens are unique individuals, as is every human being. Having a series of organisms contributes to our understanding of the range of characters within a species essential to creating robust descriptions and aiding in classification. The range of variation in characters enables the description and comparison of different sexes, ages or populations. Importantly, most have been collected from many different locations and at different times. The data on their labels is a record of the existence of an organism in space and time and together they form a time series of biodiversity over the last 500 years, and in the case of fossils hundreds of millions of years. We will discuss how that data can be used below.

The importance of demonstrating value and justifying spending

Generally the public understand the value of keeping and displaying archaeological finds, items of social history and artworks for enjoyment and records of our cultural past. Within the scope of natural science they will also appreciate the value of dinosaur skeletons, stuffed animals and other natural objects on display. People may not have the same intuitive perception of value for leeches, dung beetles, volcanic ash samples, maize seeds and minerals from the Earth's deep crust. Yet such specimens and many others have untold research value and create the broader explanatory context for the more easily appreciated cultural and iconic biological collections. If we fail to justify this spending on collections acquisition, maintenance and care, however little the cost relative to other research infrastructures such as telescope arrays or large particle accelerators, they will remain at constant risk. The number of collections orphaned by universities and other organisations are a testimony to the need for advocacy and awareness.

Example: Pulp the herbarium! An example argument for not retaining all natural science collections

The statement

There have been many arguments presented against the retention of collections and many of these often revolve around storage space which is itself a cost. In an age when storage space is at a premium, the justification for the thousands of cubic metres devoted to maintenance of collections has been questioned. A letter to *Nature* entitled "Where now for taxonomy?" (Clifford, Rogers & Dettman, 1990) posed the questions: "For well-known floras, such as those of Europe, can the retention of large numbers of herbarium sheets be justified? What would be lost if label data on all sheets was recorded, a careful selection kept and the rest pulped?" (p. 602). In essence they proposed that herbarium specimens, with the exception of types, could be digitally imaged and the original specimens destroyed thus freeing up money and space for a drastically underfunded branch of biology, taxonomy. This echoed an earlier prediction made by Paul Ehrlich (Ehrlich, 1961) that by 1970 "Large general collections will retain value as records of the diversity of life, but their value in scientific research, already much decreased, will become nil" (p. 158).

A response

Whether or not Clifford et al. (1990) had deliberately set out to provoke a debate, they certainly succeeded. A flurry of letters appeared in *Nature* and elsewhere making the case for retention of collections. The letters were not, as might be expected, only from the world's great natural history museums, but also from research scientists in a range of fields. For example:

> This demonstration project will make life harder for those who are too narrow-minded to understand the scientific value of museum specimens. For example, Clifford et al. (1990) have just offered a modest proposal to save money and improve the quality of taxonomic research, by destroying herbarium non-type specimens of plants that have recently undergone taxonomic revision
>
> (Diamond, 1990, p. 335)

The box presents an argument for disposal of collections by destruction of all but type specimens. The pressures of space and cost remain to this day, and museums funded by local government authorities struggling to fund social needs for example, might reasonably question why their ratepayers are funding rooms full of "dead stuff". Universities in need of space or experiencing changes in staff expertise and research objectives may well continue to dispose of their collections, but hopefully homes will be found elsewhere (see Chapter 6).

Exercise

Thinking about your collections:

- What arguments would you use to defend the relevance of your collections, or promote awareness, to decision makers both within and outside your organisation?
- Has disposal of a collection by whatever means been discussed in your institution? What was the outcome?

In the next section we will look at how natural science collections are proving their worth to society by supporting valuable and essential research, but before leaving the arguments against it is worth considering our own responsibilities for using what is often tax payer's money wisely. We as collections professionals may throw our hands up in horror when presented with arguments for disposal, but perhaps it also behoves us to look at how we are managing our collections. Are we doing it as efficiently as we can? Could we be more selective and focussed in how and when we collect more material (see Chapter 6)? And, particularly in the context of this book, could we manage our collections in a more strategic way with clear and evidence-backed plans?

The value of collections to society

"Value" can be an emotive word. Attempts in the 1980s and 90s to assign a monetary value to natural science collections provoked discussion within the museum community culminating in a conference on the topic in 1995 (Nudds and Pettit, 1997). There are obvious and often significant monetary values to gemstones, many fossils and tropical butterflies for example, but for most natural science objects there is no legal market. Much more interesting are the wider, societal values for study, display and inspiration of natural science collections.

In Chapter 1 we listed some of the characteristics of natural science collections which make them different from most other collections. Some, such as sheer numbers and vulnerability, are a challenge but there are many others, such as the range of temporal and geographic data held on their labels, which give them great value to a range of users from climate researchers to historians and from artists to law enforcers. For hundreds of years, natural science collections have served as tools for scientists aiding not only the description, naming and classification of the natural world but also evolutionary studies, biogeography, comparative anatomy and many more. Secondly, they have been to this day tangible, even emotive, objects for public display to excite wonderment and educate – real things to see and appreciate alongside the plethora of still and moving images and sounds of organisms displayed in the media. Here we focus mainly on research use of collections as this is where natural science collections diverge from many other museum collections.

Our collections have from the earliest beginnings formed part of a large scientific infrastructure, albeit distributed through museums and herbaria of many sizes, governances and missions around the world. Scientists researching a group of lichens generally know where to look for specimens and examine them through visits and loans to support their work through networks of colleagues. Digitisation is now making collections visible to a broader community of scientists and historians. The traditional community can also benefit from this as smaller, less well-known collections are exposing their specimens to the world.

Collections from small to large, regional to national, are tools for comparison to newly found specimens, whether by scientists to help curators assist members of the public and amateur naturalists, or for applications such as forensic science or the identification of contaminants in the environment. In addition to scientific uses natural science collections have become increasingly valuable to those researching the history of scientific endeavour. In the vast majority of these cases collections are fit for purpose without much change for these traditional uses but we may need to ask if we are delivering a service that this broader community needs and what the potential of our collections for future societal benefits is.

Uses of natural science collections

Exercise

- Thinking broadly and beyond current needs, can you list the current and potential future uses of your natural science collections?
- Would your collection be able to service these uses, and if not why not?

There is now a wealth of literature listing uses and potential uses of natural science collections, including general reviews (see Further Reading) and many papers describing research that has relied on collections to answer specific questions. An excellent review of uses of collections in the context of biodiversity conservation can be found in Museum Collections and Biodiversity Conservation (McGhie, 2019). Other papers describing and listing a variety of scientific and other uses include Mares (2009) and Bradley, Bradley, Garner and Baker (2014). We have picked a handful of examples of uses below to exemplify the range of use, and hence value, of natural science collections beyond taxonomy and its associated services such as identification.

The Society for the Preservation of Natural History Collections has drawn up a concise summary of the impact of natural science collections, "Why Collections Matter" (SPNHC). For geological collections see the Geological Society's "Statement on the Value of Museums and Collections" (Geological Society).

What are these new uses of collections?

The references above will give you a broad idea of the many uses of collections, new and potential, and here we expand on a few selected areas to emphasise how collections and societal challenges are linked.

Tracking change

The geographic information associated with each specimen is potentially valuable as baseline data for studies of change. The effects of changing climates on the distribution of species can be mapped against past distributions or used to document the spread of invasive species: a seemingly new plant to a region can be shown by examples from the collections to have been present in the past. Collections may be the only evidence of communities destroyed by deforestation and other land use changes. The value of collections as monitors of change was recognised over 100 years ago by the director of the Museum of Vertebrate Zoology of the University of California, Joseph Grinnell, who, while bemoaning the loss of native species through competition from, or hybridisation by, introduced, exotic species, noted that "It is highly desirable that a good representation of specimens of the pure, native stock be properly preserved in our museums, for future comparison." Grinnell (1910, p. 166).

Herbarium specimens often have flowers present and the label in most cases gives an exact or to the month date when collected allowing comparisons of past and present flowering times in a species. Adding to this data from collections and field observations on emerging times of pollinators builds a picture of how climate change is affecting these and its potential effects on crop plants (Robbirt, Davy, Hutchings & Roberts, 2010; Robbirt, Roberts, Hutchings & Davy, 2014).

Epidemiology

Many of the world's most infectious diseases are "zoonotic", i.e. they can be transmitted between human beings and animals. Museum collections of mammals and birds hold evidence of past infections and help unravel the history of these diseases. Drawing attention to the underuse of museum collections in disease research, DiEulii, Johnson, More and Schindel (2016) state "Reference collections such as these are our only window into the past geographic distributions of different diseases" (p. 5). They highlight the successful use of museum mammal and bird collections to map the 1918 flu outbreak and the missed opportunities to utilise natural science collections to predict outbreaks of other more recent deadly diseases such as Ebola.

Botanical specimens can also play a part in unravelling the history of diseases. Dramatic improvements in DNA sequencing make it possible to detect and identify the many fungi, viruses, bacteria, insect larvae, nematode worms and many others parasites and organisms hosted by herbarium specimens when they were living, but were very hard to identify or even detect based on morphology alone. For example DNA sequencing of potato specimens in herbaria and the pathogens within them have unravelled the biological history of the devastating potato blight that contributed to the Irish Potato Famine of the 19th century. This has contributed to our understanding of the evolution of plant pathogens and how to control them (Yoshida et al., 2013).

Species discovery

Frequently, new species are discovered not on current field expeditions but in museum collections many years after they were originally acquired. It has been estimated that only some 16% of new plant species were described within five years of collecting (Bebber et al., 2010). One quarter of species descriptions are from specimens that were collected more than 50 years ago. Your collection may well contain such unknown and undescribed species and only through expert study will they be found. It can be argued that as much resource should be used on exploring our collections as in collecting more in the field.

Historical research

Our collections play an important role in the research of historians of science and culture and provide a means to engage with these groups. They have been traditionally studied by authors investigating the lives of the "great naturalists" that created them. More recently, the historical community has found them to be a valuable resource for wider understanding of scientific cultures. Studies of the collections of Sir Hans Sloane for example have thrown light on the networks of correspondence and exchange of the 17th and 18th centuries and their importance in knowledge generation.

Balancing the books

There are many more such examples that load the scales in favour of the value and utility of natural science collections versus their cost. This balance sheet approach is eloquently presented by Suarez & Tsutsui (2004). They summarise:

> storage and maintenance of museum collections is inexpensive compared with the potential costs of their absence. Indeed, these collections confer economic benefits by serving as centralized locations for information processing and storage, saving other institutions and taxpayers, hundreds of millions of dollars per year.
>
> (Suarez & Tsutsui, 2004, p. 66)

To return to the arguments of Clifford et al. (1990), for digital capture and destruction it is clear that whilst some of the scientific value of collections can be gleaned from images alone, far more rely on physical specimens. An image lacks the testability of a specimen – a realisation that goes back to the early natural historians. Studies on changing distributions of organisms through time require accurate data and experts are increasingly able to verify the identity of a specimen from an image through morphological characters. However, molecular characters such as DNA sequences cannot be obtained from anything other than a physical specimen. Pesticide residues, chemicals and isotopes can only be found in real birds' eggs and feathers or mammal hairs, not from images. As members of the collections community we are well aware of the potential of our collections as a resource for a variety of critical studies. However, others in the museum may not be so aware. We should be prepared with current examples to illustrate the enduring value of our collections.

Do potential users know the value of your collections and how can you engage them?

If, as Suarez and Tsutsui (2004) suggest, our collections can provide an economically viable service to society, including the many uses listed above that would not have been considered 30 years ago, we can ask if we are providing a service that is fit for purpose. This is not just for now but for the future. On the whole, scientists would agree that our collections are useful, although the slow rate of digitisation, restrictions on destructive sampling and the constraints of legal and ethical agreements remain barriers to some uses. Improving both service and range of uses of your collections are strategic issues that could affect how you utilise your resources and train your staff.

Exercise

Thinking about the potential uses of your collections:

- What do you think the barriers to these uses might be and how can you overcome them?
- How can you encourage collections use among a wider audience?
- Do you know how your collections are being used now, i.e. what outcomes were there from loans and samplings from collections by external users?

How do we know how our collections are being used and what are the outcomes of this use?

Being able to demonstrate with evidence that our collections are indeed being used for worthy purposes is an important element of justifying our expenditure on collections and we will return to this again in Chapter 8 in the context of access. We can measure how many loans of specimens are made to how many institutes and we can count visitors to the collections (see Chapter 3). We will usually also know which groups of organisms they are interested in. However, do we really know what the outcomes of this research are? Scientific journals may expect authors to cite which collections they used, but unless the author sends a reprint we could be faced with the mammoth task of trawling through these journals for reference to our collection. As more and more journals are delivered in electronic form it may become possible to search for key words and identify uses that way and, as authors use unique identifiers such as ORCID IDs (https://orcid.org/), some tasks may be eased.

How can you encourage new users?

The pressure to demonstrate relevance has led holders of natural science collections to examine what they are delivering to society and to look beyond their traditional boundaries for new users. In 2008, a workshop sponsored by two European projects (the European Distributed Institute of Taxonomy [EDIT] and the SYNTHESYS project) brought together collections holders from museums, herbaria and others with representatives of a variety of actual and potential user communities including and beyond taxonomy. They listed uses, opportunities and barriers. Although some years ago, most of the opportunities and challenges identified hold true today. We have included some of the results of this workshop in Appendix 2.1. Perhaps take a look at this and see how many of these issues have been addressed in your institution or the natural science community, how many are in the process of being addressed or which you think are insurmountable? Many of these are associated with improving access to collections. Some of these relate to culture and personal interactions. If researchers see collections as disposable assets, while curators see them as precious objects, as is often the case, then that is a basis for poor interaction between new users and custodians. Our collections may be a valuable model of the living world past and present but they have some inherent weaknesses. Collections managers and researchers will have noted that their collections often have a disproportionally large numbers of specimens of rare and unusual species. Specimens of the common place daisy or house sparrow are relatively few. This temporal and geographical "patchiness" of collections reflects over 200 years of biases of collecting fashion and colonial expansion. A proportion of our specimens, particularly in collections rarely or not examined by experts, bear outdated or incorrect names adding an element of uncertainty to any study. However, when viewed as one big global collection these inconsistencies begin to even out – this is the vital role of our virtual collections – a resource that everyone can access and array the data to help answer scientific, historical and other questions. This concept of scientific collections as an array that can be purposed to address a range of new and global questions forms the basis for a number of large-scale international collaborative projects.

Crossing the boundaries and forming partnerships

There are great benefits to be gained by natural science collections and the user community from forming partnerships and collaborations, and examples range from sharing best practice to developing common access mechanisms (e.g. SYNTHESYS). Natural science collections beyond traditional museum specimens include human tissues, ice cores and many more that support scientific research and there are benefits of viewing collections together under one international umbrella. In this way, they can be arrayed to answer key questions from human origins to new emerging diseases to climate change. All specimens in collections large and small are part of one global collection in which almost every item contains unique data.

The largest ever formal agreement between collections holding organisations to date is The Distributed System of Scientific Collections (DiSSCo) (www.dissco.eu/), a European collaboration of over 100 institutions. This research infrastructure aims to bring together 1.5 billion specimens, representing 80% of all known species, online at a single website available from 2025. However it will be some years before all collections are digitised and, as we emphasised earlier, a digital surrogate does not replace the physical specimen, although it does allow researchers to use these data directly and to find where specimens of interest are housed for them to visit or loan for study. In the meantime, we can look for other means of maximising access to our physical collections through collaborations and agreements.

Collaborations can be on a variety of scales and may be project based. A project to map the seaweeds of the UK through time drew on collections from 14 museums and herbaria in the UK, from large national to smaller regional and local museums. All had seaweed collections which formed pieces of a jigsaw puzzle, and through this project all parties benefitted from public exposure of their collections via the project website (www.seaweeds.myspecies.info).

Using information on value in strategic planning

Value can play an important role in your strategic planning process and particularly in assisting you to prioritise collections for improvement. If, for instance, a particular group of insects is of interest to a wide ranging and important research project on the spread of pest species, then maybe that group should have a higher significance rating (see Chapter 3). However there is a caveat here: there is a risk that the personal interest of a researcher might lead to resources being diverted to improving a particular collection which is then found to have no value to that programme and we are left with a class A collection that is not used while others languish. Any decision should be made on the basis of wide discussions with the research and collections communities and evidence provided for the sustainability of such work.

Can we predict the future needs of our users?

We may not be able to forecast the exact needs of future users. Nonetheless, recent history tells us that users find answers to new questions in older collections and new value in those collections when new technologies are applied – and the more accessible the collection, the more likely that it will have continuing use and associated value. In a chapter about the future of collections, Norris (2018) states that "Natural history collections face four main challenges: acquiring material, preserving that material, making it available for use, and making the case that the first three activities are worthy of support" (p. 14). Norris points to numerous publications that attest to the value of collections as research infrastructure. He turns to the larger societal matter regarding access, noting that access to the discoveries and findings that result from research uses, and the associated benefits, are an essential element to a widespread perception that collections have enduring value. Norris sees digital access to collections as having a profound effect, and that is certainly true in our experience.

What other factors can influence future needs and uses? Can we predict how existing trends and possible future events might affect our collections? What can we do to ensure that decisions made now are minimally affected, i.e. future-proof our collections as much as we reasonably can. How do we balance potential future use against current needs, for example by permitting destructive sampling or by using an affordable but more invasive preservation technique?

We may look at trends in the use of our collections and speculate as to what research areas might become more important in the future. But are we best qualified to do that? Maybe this is best done in liaison with the research community who may be better placed to imagine future topics. As collections managers we can take heed of how past preservation decisions have affected the utility of today's collections. While we are increasingly able to repair DNA damaged by time and preservation methods, this would not have been necessary if other methods had been used. But hindsight is a great gift and when these specimens were collected the primary concern may have been to survive a long sea voyage or to avoid destruction by insect pests and so remain useful for morphological study. Perhaps at its simplest level the answer lies with preventative conservation of and minimum intervention with specimens, whether at point of collection or storage, i.e. to maintain the specimen in as close a condition to its living form as possible. A current example is the practice of storing leaves at the time of collection in a sealed bag with silica gel desiccant for molecular analysis alongside a traditionally heat-dried herbarium specimen from the same plant.

While we have shown how museums are making their collections more relevant to society, it may also be that they share a responsibility! The Taipei Declaration of the International Council of Museums (ICOM NATHIST, Taipei 2015) states: "Increased human activities have created catastrophic declines in biodiversity. Both ethics and logic point to a mandate to conserve vulnerable habitats and species. To achieve best practice, natural history museums take action to conserve natural habitats and populations".

How can you encourage financial or political support for your collections?

We have cited, we hope, worthy reasons for maintaining diverse collections of natural objects and have seen how on the overall "balance sheet" collections are an asset. However, this is not necessarily evident to other audiences such as potential supporters, financial and political individuals and groups that we as collections professionals interact with on an increasingly regular and necessary basis. How can we use this information to advocate for our collections and answer those questions about why we have so many flies from the same locality?

What is advocacy? In their advocacy toolkit the Natural Sciences Collections Association (NatSCA) (https://naturemanchester.wordpress.com/2014/08/13/advocating-for-natural-history-in-museums-some-tools-to-help/) defines it simply as "influencing to get results". In our case this might be funding for a new collections store, a new post or to resist further cuts to staff and collections. This resource provides support on how to promote and advocate for what we want for our collections. For example, the NatSCA advocacy toolkit identifies three main areas of advocacy: raising awareness, lobbying and collective action.

Advocacy can be at many levels. In a multidisciplinary museum it might be to gain a larger share of the budget for natural sciences, or at least to prevent its decline relative to other museum activities. Increasingly, as curators we are asked to meet funders and politicians and this is our opportunity to put our case for why they should give us money, political backing, etc.

There are a number of resources available that advocate for natural science collections. The Emerging Professionals Group of the Society for the Preservation of Natural History Collections (SPNHC) posted a thought provoking article, "Stealing the Spotlight in Natural History Museums: How well do we advocate for our collections in exhibits?", which highlights that while exhibits can play an important role in advocacy, the degree to which this happens varies greatly from one group of organisms to another (Corriveau, 2014).

Note that many curators and their institutions promote what they hold in their collections and, crucially, what they do with them using social media.

An important first question in advocating to politicians and funders is simply "What is in it for them?" It is easy as curators to assume that everyone should see how important their work and collections are, but this is a poor assumption. Large funding bodies will have criteria governing who and what they will support and it is relatively straightforward to match what our collections can offer to their mission. Knowing what the interest and ethos of potential donors are can help to tailor advocacy materials and "pitch" to that audience.

Exercise

Prepare a short "elevator pitch" to a potential funder to support a real or potential collections improvement project that you are planning. Why should they give you money? What is special about your collection?

For local audiences, anything that raises the profile of their town, city or region might be of interest. This may be a high profile specimen or collections associated with well-known figures – someone that puts your town on the map. This is the so called "wow factor". The presence of iconic specimens collected by Darwin or Von Humboldt are an obvious draw but maybe the local value is more subtle. As well as the global aspirations of climate research, collections are a record of the *local* past and can be integrated into school studies on their local changing landscape. This use is of particular value in displays and education, but also when advocating on a local level, and might include specific problems or threats to the local environment, objects around which local stories can be told. Having a good understanding of the history of natural science and collecting provides an excellent basis for story-telling and provides hooks to engage audiences on the importance of natural science today. It is also an area of common ground with other cultural collections and understood by the public and potential donors, funders and politicians. Identifying specimens in your collections of specific historical interest, whether collected by "celebrities" or as evidence of cultural changes such as colonization and empire building, is an important tool for advocacy and might also influence prioritisation for improvements to those collections as part of your planning process.

The bottom line is to have as much intelligence as can be gleaned on the target audience and ensure that you have clear facts and are not overselling. Saving the world is not always an obvious reason!

As we have discussed, there are many and varied organisations and individuals who have or could have interest in our collections. To maximise our services to the community we need to understand who these stakeholders are and how strong our relationship is. Such knowledge will help you with planning, prioritisation and focus of collections services.

There are a growing number of websites that can act as a showcase for the collections held in any museum, herbarium or university. These provide data at a broad level often as simple as "we have lichens" or "our collections are mainly from region X". These websites are searchable and not only provide simple means for researchers to locate what they need but can be used in advocacy to demonstrate how all collections are important. For example the Global Biodiversity Information Facility (GBIF) now includes a searchable registry of biological collections, GRSciColl (www.gbif.org/grscicoll).

Finally, you cannot underestimate the value of dialogue with your stakeholders and also with potential donors/funders rather than a one-sided hard sell. Try to understand their needs and wishes, try to get them engaged in the process and not (always) try to find answers by yourself. Some of your stakeholders might be your greatest advocates!

Summary

Understanding the history and context of your collections and realising their potential for current and future use has a number of strategic values and advantages particularly in advocacy for collections management work but also in informing planning decisions for future work. We will develop these themes further in the following chapters. Equally important is the recognition that the cost of scientific collections is outweighed actually and potentially by their value to society. As collections managers we have a responsibility to communicate this message to our funders, politicians and also potential users. This is a professional responsibility. We can use stories built around the use and value of our collections to advocate for their retention, preservation and use. Some collections will be relevant to particular scientific questions and research and this value may influence which collections we take the strategic decision to improve before others (see Chapter 3) or perhaps to digitise first (see Chapter 7). To ensure we have these resources available for the future we may need more collections staff with a broad range of skills and knowledge to develop and apply advocacy strategies, to engage the research community to match the resource with the needs of the users and to build networks to maximise ability to answer a wide range of questions. This must be in the context of long-term preservation – our collections must be fit for purpose not just for today but long into the future. The role of social media in these endeavours cannot be overestimated and there are few museums and herbaria that are not using these to their best advantage. The message should be "Look what we have got and this is what you can use it for and this is why it is important to our society".

It is clear that we are facing significant changes in natural science collections and how they sit within our society. Whether due to the "push" of societal challenges or the "pull" of emerging new technologies, these will change how we manage our collections and how we will provide access. These changes will build on the digital transformation of our collections and include much stronger engagement with the general public, whether through citizen science or displaying what we have to offer to society through our exhibitions and educational programmes. We have a framework given to us and it is our task to develop a strategy to either adapt or change our ways of working, and it is our responsibility to take care that the collection is ready to answer new needs but also safeguard our collections through rapidly changing times. We need to find the right balance between sustainable preservation and staying relevant.

Notes

1 When first founded, The British Museum in Bloomsbury London held natural science objects as well as antiquities. In 1881 the natural science collections were rehoused in a new purpose built building in South Kensington, now known as the Natural History Museum.

2 In botanical and mycological collections there may be several specimens made from one organism. For example branches of the same tree each preserved with a label. These are known as "duplicates" – samples from a single individual specimen and hence genetically similar. Duplicates are created and distributed to several institutions to enrich their collections.

References

Asma, S.T., 2001, *Stuffed Animals & Pickled Heads: The Culture and Evolution of Natural History Museums*. Oxford: Oxford University Press.

Bebber, D.P. et al., 2010, Herbaria Are a Major Frontier For Species Discovery. *Proceedings of the National Academy of Sciences*, 107 (51): 22169–22171. (doi: 10.1073/pnas.1011841108) www.pnas.org/cgi/doi/10.1073/pnas.1011841108 [Accessed 21/10/19]

Bradley, R.D., Bradley, L.C., Garner, H.J, and Baker, R.J., 2014, Assessing the Value of Natural History Collections and Addressing Issues Regarding Long-Term Growth and Care. *Bioscience,* 64 (12): 1150–1158. https://doi.org/10.1093/biosci/biu166 [Accessed 8/10/2019]

Clifford, H.T., Rogers, R.W., and Dettman, M.E., 1990, Where now for Taxonomy? *Nature,* 346: 602.

Corriveau, K., 2014, Stealing the Spotlight in Natural History Museums: How Well Do We Advocate for Our Collections In Exhibits? https://crackingthecollections.wordpress.com/2014/05/10/stealing-the-spotlight/ [Accessed 06/11/19]

Diamond, J.M., 1990, Old Dead Rats Are Valuable. *Nature,* 347: 334–335.

DiEuliis, D., Johnson, K.R., More, S.S., and Schindel, D.E., 2016, Opinion: Specimen Collections Should Have a Much Bigger Role In Infectious Disease Research and Response. *PNAS*, 113 (1): 4–7. https://doi.org/10.1073/pnas.1522680112. [Accessed 06/11/19]

Ehrlich, P.R., 1961, Systematics in 1970: Some Unpopular Predictions. *Systematic Zoology*, 10: 157–158.

Freeman, P., 2011, *Alexander the Great.* New York: Simon and Schuster.

Geology Society. Statement on the Value of Museums and Collections. www.geolsoc.org.uk/~/media/shared/documents/policy/Statement%20on%20Museums%20and%20Collections_Final.pdf?la=en [Accessed 21/11/19]

Grinnell, J., 1910, The Methods and Uses of a Research Museum. *Popular Science Monthly*, 77: 163–169.

Hykin, M., Bi, K., and McGuire, J.A., 2015, Fixing Formalin: A Method to Recover Genomic-Scale DNA Sequence Data from Formalin-Fixed Museum Specimens Using High-Throughput Sequencing. *PLoS ONE*, 10: e0141579. (doi:10.1371/journal.pone.0141579) [Accessed 3/11/19]

ICOM NATHIST, Taipei, 2015, https://icomnathist.wordpress.com/taipei-declaration-on-nhms-and-biodiversity-conservation/ [Accessed 06/11/19]

Mares, M.A., 2009, Natural Science Collections: America's Irreplaceable Resource. *BioScience, 59*: 544–545.

McGhie, H.A., 2019, *Museum Collections and Biodiversity Conservation.* Curating Tomorrow, UK. www.curatingtomorrow.co.uk [Accessed 06/11/19]

Norris, C.A., 2018, The Future of Natural History Collections. In: E. Dorfman (ed.), *The Future of Natural History Museums* (pp. 13–28). Oxon and New York: Routledge.

Nudds, J.R., and Pettit, C.W. (eds.), 1997, *The Value and Valuation of Natural Science Collections*. London: Geological Society.

Robbirt, K.M., Davy. A.J., Hutchings, M.J., and Roberts, D.L., 2010, Validation of Biological Collections As a Source of Phenological Data for Use In Climate

Change Studies: A Case Study With the Orchid Ophrys sphegodes. *Journal of Ecology*, 99, 235–241.

Robbirt, K.M., Roberts, D.L., Hutchings, M.J., and Davy, A.J., 2014, Potential Disruption of Pollination in a Sexually Deceptive Orchid by Climatic Change. *Current Biology*, 24 (23), 2845–2849.

SPNHC. Why Collections Matter. https://spnhc.org/what-spnhc-does/why-collections-matter/ [Accessed 21/11/19]

Suarez, A.V. , and Tsutsui, N.D., 2004, The Value of Museum Collections for Research and Society. *Bioscience*, 54 (1): 66–73.

Woodward, J., 1728, *An Attempt Towards the Natural History of the Fossils of England*. In J. Woodward, *A Catalogue of the English Fossils in the Collection of J. Woodward*. London: F. Fayram.

Yoshida, K. et al., 2013, The Rise and Fall of the *Phytophthora infestans* Lineage That Triggered the Irish Potato Famine. *Elife*, May 28, 2: e00731. (doi: 10.7554/eLife.00731. Erratum in: Elife, 2013, 2 : e01108). [Accessed 3/11/19]

Further reading

Huxley, R., (ed.), 2019, *The Great Naturalists* (2nd edition). London: Thames & Hudson.

Heberling, J.M., Prather, L.A., and Tonsor, S.J, 2019, The Changing Uses of Herbarium Data in an Era of Global Change: An Overview Using Automated Content Analysis, *Bioscience*, 69 (10): 812–822. https://doi.org/10.1093/biosci/biz094 [Accessed 20/11/2019]

Kemp, C., 2015, The Endangered Dead. *Nature*. 518: 292–294. www.nature.com/news/museums-the-endangered-dead-1.16942

Lavoie, C., 2013, Biological Collections In An Ever Changing World: Herbaria As Tools For Biogeographical and Environmental Studies. *Perspectives in Plant Ecology, Evolution and Systematics*, 15: 68–76.

MacGregor, A., 2018, *Naturalists In the Field : Collecting, Recording and Preserving the Natural World From the Fifteenth To the Twenty-First Century*. Leiden: Brill.

Pyke, G.H., and Ehrlich, P.R., 2009, Biological Collections and Ecological/Environmental Research: A Review, Some Observations and a Look To the Future. *Biological Reviews*, 85 (2): 247–266.

Schindel D.E., and Cook J.A., 2018, The Next Generation of Natural History Collections. *PLoS Biol*, 16 (7): e2006125. https://doi.org/10.1371/journal.pbio.2006125 [Accessed 1/11/2019]

Tewksbury, J.J. et al., 2014, Natural History's Place In Science and Society. *Bioscience*, 64 (4): 300–310. https://doi.org/10.1093/biosci/biu032 [Accessed 21/11/2019]

APPENDIX 2.1: BROADENING THE USER BASE

Adapted and edited from an unpublished report on the workshop – Broadening the User Base 4–5th November 2008, Brussels. Report compiled by S. Owens and D. Duin.

The workshop was a joint activity funded through the European Commission SYNTHESYS Research Infrastructures programme, under FP6 called "Synthesis of Systematic Resources", and the European Distributed Institute of Taxonomy (EDIT). EDIT funded as a "network of excellence" (NoE) by the European Commission (6th framework programme, FP6). This report reflects the views only of the authors. The Commission cannot be held responsible for any use which may be made of the information contained therein.

Uses of natural history collections

- Environmental specimen banking for environmental monitoring.
- DNA samples for: archaeology, wildlife forensics, customs, other.
- Long trial studies in population genetics.
- Environmental impact assessments, e.g. for landscape planning.
- Natural history art exhibitions, books.
- Long term biodiversity modelling.
- Agriculture – identification of biological control agents for weeds and plant diseases.
- Human and animal health – specimens as reference collections for disease factors.
- Identification of wood, e.g. antique furniture, detecting illegal logging.
- Inspiration for engineering products (biomimicry).
- Education of the general public on natural history, biodiversity and evolution.
- Identification and naming of pest organisms especially those belonging to species complexes (food, agriculture, human and veterinary health, environmental studies, policy makers).
- Prediction of changes in distribution of organisms through global change, warming or land use (food, agriculture, human and veterinary health, environmental studies, policy makers).
- Prediction of the potential damage caused by invasive pests and vectors of plant diseases (food, agriculture, human and veterinary health, policy makers).
- Increased global knowledge on quarantine organisms (customs services, food, agriculture, human and veterinary health, policy makers).

Needs of users

- Reliable and accessible database records.
- Georeferenced database records.
- Appropriate, flexible institutional conditions to respond to user demands.
- Web-accessible electronic keys for determination/identification of species by non-taxonomists.

- Access to specimen collections including rare specimens.
- Access to "paper" archives, such as books, maps, drawings, personal correspondence.
- Access to digital images of specimen collections and paper archives, 3D scans.
- Clear contact point for user enquiries on website (name, telephone and email).
- Accurate records (coordinates, rather than provinces or ecological regions).
- A funding mechanism to assist access to users from outside the taxonomic community.
- Institutional support for making a broad user base a priority.
- Non-conventional users additional to conventional users, not instead of.
- Infrastructure could charge users for certain services, under certain conditions.
- Additional staff to look after user requests.
- Training for staff to be able to better respond to requests.
- Infrastructural support; provision of communication services, work space and equipment for different types of visitors.
- Mechanisms to allow destructive sampling (and minimise damage).
- More taxidermy expertise available.
- Flexible and plentiful storage space in particular frozen storage.
- Platform to consult users regularly on their needs.
- Solve legal issues regarding copyright, liability insurance for visitors etc. that hamper use and access.

Barriers to broadening use

- Little incentive (job grading) for collection managers and research staff to work with non-conventional users.
- Lack of staff to reply to all user requests.
- Lack of coordination at international level, e.g. collection standards.
- Lack of long-term preservation of (database) technology.
- Scattered collections in several countries with no clear point of contact, and hard to find what collections are where.
- Limited amount of material available from traditional specimens, e.g. for DNA extraction.
- No overview of current state of collection use. Difficult to monitor use of collections, e.g. in publications, no systematic coordinated way of referencing to the loan institutions.

Proposed mechanisms to broaden user base

- Develop mechanisms to monitor use of collections.
- Train staff to deal with user requests.

- Create funding mechanism to support users from countries of origin.
- Have a clear communication strategy towards users on the websites of institutions and proactively in newsletters and meetings.
- Global networks of collections, e.g. GBRCN's and the International Federation/Consortium of Scientific Collections from the OECD GSF will make it easier for users to communicate with natural history collections.
- Collections managers should participate in research user networks like EPSO (European Plant Science Organisation).

3 Strategy and management

Introduction

In the previous chapter we explored the current and future users of scientific collections and what they require from a collection. Few, if any, scientific collections are perfect for the needs of their users. Some may be in too poor a state of preservation to be handled, whereas others might be organised in outdated systems making locating collections difficult and making them less likely to be accessed. Identifying and addressing risks and prioritising remedial action requires careful planning at a strategic level, whether a collection is comprised of a thousand mosses or 5 million beetles. Strategic management of collections depends on a series of planned, deliberate actions that connect all areas of collections work and drive the organisation towards achieving its purposes. This is true regardless of the kind or size of collections-holding organisation, whether it is a research institute, university museum, cultural centre, regional or state museum, or national museum. Strategic management and the actions that support it do not typically occur in a vacuum: they are the result of plans. Planning requires a clear understanding of the purpose of the museum, the purpose or intellectual framework for the collections, the interests and needs of users of the collections and the resources available to the museum. The best plans are developed using current data and information, and they contain clear action steps that can be reviewed for their effectiveness. Strategic plans for collections consider one of the fundamental tensions in museums, which is the perennial balancing of preservation with use. Good strategic collections management enables a smoothly operating museum, a productive staff and collections that are well preserved, easily accessed, well used and well prepared to meet the needs of future users.

In this chapter we explore strategic planning and examine some of the ways that you can gather information to develop a plan and monitor progress towards objectives. As you read, remember that an effective strategic plan is based on agreed needs and priorities, supported with adequate resources, with actions that can be measured against goals for progress.

Questions

- What is a collections strategic plan and how will it help me?
- How does strategic planning relate to the purpose, values and policies of my museum, department or project?
- How can I get information or data on the current state of my collections to identify collections needs and help with planning my project?
- What key elements do I need to include in my strategic plan?
- How can I make the planning process easier?
- How can I prioritise the actions in my plan?
- Are there templates that I can use to organise my collections actions?
- How can I know that I am meeting my collections objectives in the longer term?

What is a collections strategy and how will it help me?

Strategy is a way of expressing what the museum, department or project will do to achieve its purpose and vision. For a museum that holds collections that are not accessible via digital means, a digital strategy might be to "make the collections accessible via our website". Tactics are the action steps the museum will take to accomplish the strategy. Following the example above, tactics might include creating digital records for selected specimens, improving the search engine on the museum's website and making metadata available via national data portals.

Collections management strategy is integral to achieving the purpose of the museum. Managing collections requires specific strategies that intersect with strategies for exhibition, education and research, and are consistent with collections preservation and operational considerations such as budget, finance, facility management and personnel management. Museums typically have higher level strategic plans that guide their work, and sometimes make these available on their websites, and if not, most will make copies of their plans available upon request. These plans are at relatively high level to the plans you will be drawing up to achieve your own collections goals and you may need to ensure that there is no conflict. Plans can be organised in different ways but they usually cover topics such as:

- what is to be acquired and disposed of – Collections Development Plan;
- the amount and nature of the space for collections housing, processing and use – Space Plan;

- representing the collections and information in digital format – Digitisation Plan;
- conservation, rehousing, organisation – Preventive Conservation or Physical Care Plan;
- exhibitions and education, communicating relevance, training materials, communication on use – Communications Plan or Public Interpretation Plan;
- unifying plan covering all aspects of the museum (usually time based) – Strategic Plan for year n – year n+.

In some organisations, the first four of these might be wrapped up into a general collections plan.

The principles of strategic planning can be applied at all levels. For example, there can be a strategic objective to ensure that all those collections amassed for molecular analysis are accessible to a wide audience. The tactics to achieve this might be to locate, count and digitally record data on these collections. A more ambitious plan could be to create a purpose-built storage facility for them. Plans require a baseline as a starting point. The manager in charge and the team will need to articulate the nature of the collection, what it contains and why it exists – this information is referred to as an intellectual framework.

Strategies for collections, but also for other disciplines such as public engagement, are built using interpreted data that is viewed in light of the purpose of the museum and, in this case, the intellectual framework for the collections (see below). They are based on understanding the current state of the collections and the desired future state. Tactics or action steps move the collections towards the desired state. One way that museums gain information for these plans is by conducting assessments. Assessment should lead to the development of management systems, not simply printed reports that are ignored. The importance of assessments at a useful level of detail that reveal objective data cannot be stressed enough and we devote a large portion of this chapter to that topic.

Benefits of strategic plans are both general and specific. In general, they emerge from broad consultation across stakeholders and experts and thus reflect a shared viewpoint regarding the museum. They also reinforce the purpose and values of the museum. Specifically, strategic plans guide resource decisions and work plans, and can help resolve controversies about what actions or opportunities will not be pursued. Evidence of strategic planning can be effective when seeking external funding or pursuing accreditation: in fact, the presence of a strategic plan is sometimes necessary for gaining donated funds, so well-crafted plans can have highly beneficial results for the museum. They can also help focus work so that collections are improved and are more readily used. This enhances the likelihood that a collection maintains its relevance and long-term support.

Exercise

Consider your collections:

- Do you know the mission and vision of your museum?
- Is there a current museum-level strategy?
- How does this strategy relate to the management of the collections under your care?
- Where might you benefit from strategic planning?
- What are the benefits you already observe from strategic plans your museum has developed?

Benefits of strategic plans

Perhaps the most significant result of successful strategic planning is the realisation of the museum's vision and strategic goals, and a well preserved, well developed, accessible and representative collection. In some instances, successful planning may be acknowledged by gaining accreditation or other formally recognised certification that professional standards are achieved. Other results can be timely accomplishment of goals such as planned preventive care measures, construction of new facilities, relocations of collections to improved spaces, strengthening an existing collection through planned acquisitions and so on. Strategic plans can also result in the museum gaining funds for goals in the plan, or in the museum modifying the culture of the staff by reinforcing alignment with its purpose and rallying around an agreed goal. Development of a strategic plan is a way of documenting an orderly approach towards reaching a goal.

How does strategic planning relate to the purpose, values and policies of my museum, department or project?

Strategic plans reveal what is important to the museum because they focus on the phenomena or topics that are essential to fulfilling its mission. In other words, external reviewers, as well as staff, can tell what is valued by the museum by the topics, items and problems that receive attention and resources. Strategic plans implement the museum's policies on, for example, preventive conservation by pursuing improved storage environments and organisation, better housing materials, and prioritised treatments. They allow the museum to affirmatively state that it is a responsible steward of significant, rare or valuable items because the strategic plan for care of those items is well informed, deliberate and effective. Strategic plans also draw upon the intellectual framework for the collection, as discussed below.

What is an intellectual framework for collections?

The AAM Guide to Collections Planning (Gardner and Merritt, 2004) states that "an intellectual framework is the key to establishing intellectual control over the museum's collections" (p. 5). It articulates the rationale for collections and collecting. It states the compelling vision that defines the museum's role and provides the context for making decisions about the future of the collections. In essence, the intellectual framework articulates what we have got and why we have it.

Exercise

Think about your collection (or maybe a subset of that collection) and its current and future use. Write responses to the following:

- Why does your collection exist?
- What preservation methods are represented in the collection (e.g. jars of liquid-preserved specimens, pinned insects, frozen tissues), or what taxonomic or stratigraphic groups are represented?
- Who are the primary users of the collection? What other types of users should be accommodated?
- How does the collection need to change to better meet the needs of the users?

Responses to questions such as those listed in the exercise above form the intellectual framework for your collection and give you a starting point to plan any changes or developments that may be needed. See below for an example intellectual framework.

Example: Hypothetical intellectual framework – Institute X

The mission of Institute X is to understand the biodiversity of Country X. Institute X's scientists study the biogeography of the country and adjacent regions, identify patterns of biodiversity and develop and test hypotheses to understand and explain the processes that shape Country X's natural environment. It is necessary to assemble plant and animal specimens to do this work. The resulting collections are necessary for both for the studies being made by the scientist who collected the specimens and also for future researchers who consult the collection as reference material to complement newly-made collections. Significant investment of resources and application of scientific expertise is made in acquiring collections. Because the collection is intended to serve as yet unidentified user audiences and research questions, the investment in the collections is sustained by managing the collections carefully and deliberately so that they are ready for future use.

Specimens have been acquired since 1900, taking the form of fluid preserved, dry and samples preserved at cold temperatures (4 °C and -80°C). Institute X intends to maintain the collection indefinitely so that future comparative studies may be made.

Institute X's collection exists primarily for scientific research. While it may be used for some educational programming, specimens are not intended for significant use in a classroom, museum gallery or other public venue. The collections are accessible via the publications that result from research, and loans may be made to researchers at other institutions. Metadata for selected portions of the collections are publicly accessible via the internet. The collections are not available for commercialisation.

Example: Hypothetical intellectual framework – multidisciplinary Museum Y

Museum Y is a beloved local museum with strong ties within the community, known for its exhibitions and educational programmes. Its original collection is comprised of approximately 500 local works of art. Recently, it has expanded its collection scope to include approximately 5,000 biological, geological and invertebrate fossil specimens and ten vertebrate fossils from the surrounding region. The natural history collections were developed by the local university as a result of professors teaching field collecting methods. An agreement by the university and Museum Y transferred the collection so that it might be used for exhibition and educational purposes, especially for public understanding the biodiversity and heritage of the region. Museum Y intends to develop all the collection for permanent and temporary exhibitions. The collection needs to represent the flora and fauna of the region, the major geological features and their fossil-rich strata, and the breadth of local artistic traditions, as they are unique in their use of biological motifs.

Intellectual frameworks are independent of scale and governance; they can describe a large national research collection or small collections used primarily for public engagement. Notice that the examples above state why the organisation and the collections exist. They clarify the intended users, and the kind of objects and specimens in the collection, state how long they will be maintained and state the factors that guide how the collection may be used. The relationship between the purpose of Institute X and Museum Y and their respective collections is clear. Because the intellectual framework states that the collection is for research, Institute X could make a compelling justification for investing resources to make the collection accessible to a broad audience of research users via digital means while continuing to preserve the specimens themselves. Collections professionals at Institute X would likely develop strategic plans that prioritise improvement projects based upon the needs and requests of the researcher community. Museum Y can draw a clear connection between their mission of presenting local heritage and their acquisition plans. Even these brief intellectual frameworks can guide development of strategic plans.

What is the desired state of the collection?

Starting from the purpose of the collection as expressed in the intellectual framework, the next step in strategic planning is to briefly clarify the desired state of the collections. When considering this, think about the needs of the stakeholders and collection users, any obligations based on the nature of the museum (e.g. state or national repository) and a realistic understanding of resources including space, funding, staff expertise and general infrastructure. Should all objects and specimens be preserved indefinitely, or can some be designated for consumptive uses? Does the existing collection sufficiently support contemporary analytical methods, or are additional objects and specimens needed? Does the collection contain objects appropriate for exhibition, or are significant conservation repairs required or newly collected specimens needed? Will you create a digital record for every single specimen or will you allow a basic record to represent multiple specimens or items? Will the collection be organised to a highly specific degree, or is a general organisation acceptable?

Discussions about the desired state of the collection provide an excellent opportunity for many different viewpoints to be considered and accommodated. Staff with different areas of responsibility and expertise should be represented in these discussions so that everyone has the same understanding of the goal for the collections. In small museums, this might mean a discussion among all staff as each person has a stake in the readiness of the collection. The director wishes to gain accreditation. The curator wishes to refresh aging exhibitions with aesthetically appealing specimens and to be able to present a temporary exhibition every year while pursuing on-going research projects with external colleagues. The educator needs teaching specimens. The collections manager is striving for conditions that promote preventive conservation while maximising digital accessibility. Regardless of the type or size of the museum or collection, inclusive and broad discussions of the desired state of the collection are essential. It is also an appropriate time to thoroughly discuss the tension between preservation and use, and how the collections are integral to supporting the larger purpose of the organisation. Once the desired state is defined, you can turn to the next steps in developing the strategic plan.

Exercise

Consider your collection. What are three statements you can make about its desired state?

Using assessment data to develop a strategic plan

Once the intellectual framework is clear and the desired state for the collections is known, the question is whether the current collection meets the desired state. An assessment or survey of the collection is a sensible approach to provide you with an overview. Most likely you will identify areas of strength or a good fit with the desired state, and areas of challenge or weakness, representing a gap between the current reality and the desired state. As achievable strategies rely on current, accurate information about realities, once the assessment data are analysed you can identify the strategic actions needed to achieve the desired state. Considering the points about relevance and users discussed in Chapter 2, we might ask:

- Do we have the depth and breadth of collections we need to serve our users and achieve our purposes?
- Are specific portions of the collection ready to use for research, education and exhibition?
- Do we understand the needs and interests of our users and audience?
- What do we know about our collections?
- Is the information about collections and individual objects available and accessible?
- Do we have enough space?
- What is acceptable space? What is high quality space?
- Do we fulfil legal and ethical requirements? Are necessary policies and procedures in place?

The desired state for a collection might be "all digital records for the rocks and ores should have georeferenced data". Or, "all cultural items from sub-Saharan Africa must have provenance data that was clarified through research and available to users in catalogue records". Or, "all new biological vouchers must be accompanied by a frozen sample appropriate for molecular analysis".

How can I get data on the current state of my collection to help with planning?

Gathering data can be overwhelming, and without a defined process it is easy to be inaccurate or inefficient. Following a defined process that uses a validated assessment schema or tool is more likely to yield objective information so that action steps can be established and prioritised than would be achieved through an unplanned process. There are a number of schemas that address different aspects of collections and museum management (see discussion below). Some staff may feel that it is acceptable to rely upon impression and anecdotes about the current state of the organization's collections. Using an "impression and anecdote" process at the onset of strategic planning may seem fast and simple but ultimately it will be costly in terms of reproducibility and acceptance of results. Instead, invest the time at the beginning of your strategic planning process to be clear on what you are assessing, the assessment criteria and the units of analysis so that your results will help you develop plans to reach the desired state of the collections. You may choose to develop a strategic plan that addresses all components of the desired state you described or just one or two. Whatever the focus of your strategic plan, you will find that objective data are essential for gaining resources and support, and for showing that your action steps achieved the desired result. Strategic plans should be based on understanding your current situation and your desired state. Plans draw upon objective data from assessments. Data alone will not allow development of a strategic plan. Instead, the interpretation of data is required. It is important to be clear about the aim and scope of the assessment or survey before starting to gather data. The following questions may help you define the requirements for the assessment:

- What aspects of collections management do you want to improve?
- What schema or tools are appropriate for the improvements you wish to make?
- How can you interpret your data to support prioritization and planning?
- Does the information you collect allow you to measure your progress over time?
- How can you demonstrate your progress?

Embarking on assessments can be potentially costly and time consuming, and you may need to convince others that they are worthwhile or influence the decision about the kind of assessment to use.

Exercise

Think about your collection. What would be the value of doing an assessment?

There are many possible assessments that one could do, ranging from gaining a general picture of how well an institution or department is performing, e.g. whether it has a disaster recovery plan, to the more specific, e.g. how many collection units (drawers, boxes, jars, etc.) are at an acceptable/usable standard. Any collection, regardless of size, type or status can benefit from assessment.

The reasons for doing assessments include to:

- determine the condition of the collection;
- reveal information on various qualities of collections;
- set the baseline for tracking improvements and accomplishments;
- provide objective data to help make decisions and plan improvement projects;
- contribute to planning for new facilities such as laboratories and specialised storage areas;
- allow more credible cost/benefit analyses and support funding requests;
- provide a summary/visualisation for non-collections staff for information and advocacy purposes;
- move collections management from anecdotal to the documented;
- contribute to performance indicators in collections management.

Types of assessments

Our purpose in this section is not to discuss every assessment method available today for understanding and improving the status of a collection. Instead, we will explore broad categories of assessments supported by some examples. Our aim is to help you explore different types of assessments and gain confidence in using them to advance your strategic goals.

The various assessments available can be categorised according to their level of granularity and also by the audience they serve or the problem they are designed to explore.

Self-assessment tools

There are a number of assessment types that measure the performance of an institution against a set of standards or benchmarks and can be performed by the institution itself without external review or input.

The SYNTHESYS Collections Self-Assessment Tool (CSAT) for example, provides a strategic overview of direct and indirect risks to a museum's/institution's collections. CSAT builds upon previous more intensive SYNTHESYS collections surveys by a panel of external experts (Collins et al., 2006). While these were very useful to the institutions surveyed (many used the results to improve internal procedures), they were costly and labour intensive. CSAT was developed as a free simplified tool that would

Table 3.1 SYNTHESYS Collections Self-Assessment Tool (CSAT) showing weightings applied to benchmarks

Risk rating	*Not met*	*Partially met*	*Mainly met*	*Met*	*Exceeded*
1	Amber	Amber	Green	Green	Blue
2	Red	Amber	Amber	Green	Blue
3	Red	Red	Amber	Green	Blue

Source: SYNTHESYS was funded by the European Commission infrastructures programme under the FP7 framework

allow more institutions to participate and would gather data on the state of collections management initially in European institutions.

CSAT (www.synthesys.info/network-activities/synthesys3-na2/self-assessment.html) is a web-based system that allows institutions to assess and record whether they meet, mainly meet, partially meet or do not meet a series of benchmarks covering all areas of museum activity that impact on collections from finances to use of correct materials for storage. The assessment software applies weightings depending on the criticality of a benchmark and returns a report with a highlighting system identifying any urgent needs for action. Failing to meet a benchmark relating to water-tightness of a building for example will receive a "red" rating (see Table 3.1) even if "partly met", whereas failure to have a mission statement would return an amber at maximum. Any benchmarks registering as red or amber are supported by suggestions for improvement.

The survey is divided into sections covering infrastructure – management, infrastructure – buildings, collections management, collections care and preparation, access to collections, molecular collections, and digital collections. Institutions have used the results to support operational changes and infrastructure investments. We have included the list of benchmarks (see Table 3.3 in Appendix 3.1. One advantage of having a standard set of benchmarks is the ability to reassess and map changes through time. Table 3.2 gives an example of a collection that has used reports from the CSAT to identify areas of priority and then taken action to improve them.

The audience are primarily the institutions themselves, but by putting all results together the project was able to identify any common areas of weaknesses across European collections and suggest ways to address them.

The SYNTHESYS assessment gives a valuable overview but does not provide detailed data on the collections, which may be needed to plan a specific project in detail. For example how many drawers of insects are in sub-optimal storage or how many fossils are housed in inappropriate climatic conditions? There have been many assessment methods devised for a variety of collections that drill down to a detailed level and we will examine these below.

Table 3.2 Combined reports from assessments carried out in 2015 and 2018 using the SYNTHESYS Collections Self-Assessment Tool (CSAT) before and after improvements were made

Benchmark	*Rating 2015*	*Criticality 2015*	*Rating 2018*	*Criticality 2018*
Repairs are made to the collections buildings as a matter of priority to ensure the collections are protected.	Partially met	RED	Mainly met	AMBER
Resources and a budget for the maintenance of collections buildings and associated equipment have been allocated.	Met	GREEN	Met	GREEN
Collections buildings and associated equipment are inspected on a regular basis and potential risks to collections are identified and reports maintained.	Partially met	AMBER	Met	GREEN
Procedures are in place to ensure that actions are taken to mitigate any risks identified.	Partially met	RED	Met	GREEN
Trained maintenance staff monitor all buildings and facilities and have sufficient budget to ensure buildings are maintained to an agreed standard and operate to ensure the environment stays within agreed parameters.	Partially met	AMBER	Mainly met	AMBER
Reports on the condition of the building and conservation advice are used in planning improvements or major developments.	Met	GREEN	Met	GREEN
Access points to the collections building such as doors, lifts, windows and ventilation are secure.	Partially met	RED	Met	GREEN
Collections are kept in a secure environment, including those that are on display/exhibition.	Partially met	RED	Mainly met	AMBER
Written security procedures and guidelines are enforced and communicated to all staff, including maintenance staff and contractors.	Partially met	AMBER	Met	GREEN
All collection and exhibition spaces are regularly cleaned by staff following sets of institutionally-agreed procedures and methods that minimise risks to the collections.	Partially met	AMBER	Met	GREEN

Table 3.2 Cont.

Benchmark	*Rating 2015*	*Criticality 2015*	*Rating 2018*	*Criticality 2018*
Environment reports (including integrated pest management) are used by those managing collections storage buildings to ensure that risks to the collections are minimised when any improvements or repairs are made.	Partially met	AMBER	Met	GREEN
There is a written emergency preparedness plan for the collections which includes procedures for dealing with disasters and the salvage of collections. All staff are aware of it and know their role within it.	Partially met	AMBER	Partially met	AMBER

Source: SYNTHESYS was funded by the European Commission infrastructures programme under the FP7 framework

The SYNTHESYS self-assessment system uses a risk factor to determine which of a series of statements regarding a collection or institution require immediate action or can be postponed or ignored. Although it might not appear obvious, nearly all activities of a museum's work can be described in terms of risk. Consider this: is there a risk associated with not having a mission statement? The effect of having no mission statement presents a risk through the indirect effects. No mission means no clear goal means no focus on preventative conservation for example. Risks and their management and mitigation are ultimately at the core of collections care and accessibility. In order to be able to prioritise which specimens, collections, buildings and so on are most at risk from agents of deterioration requires a good understanding of risk. Risk can at its simplest be defined as the effect of a particular agent, such as fire, on a collection multiplied by the likelihood of its occurrence. Simply applying a risk value to a collections store, cabinet or building can assist collections managers to appreciate what the objective, rather than imagined, risks are to their collections and allocate scarce resources effectively. There are a number of methodologies available for assessing risks to collections at a variety of levels. The Share East Museum network in the UK have produced a "how to guide" for assessing and managing risks in collections care (Hillhouse, 2012). For many museums, this system is an adequate broad tool to assist in decision making, however, in reality risk is a more complex issue. For example you might want to take into account the loss in value of your collection that a destructive event might cause – losing one leaf of many in a plant specimen could be considered less important than losing the only flower and consequently a host of characters essential to its identification and classification. There is a wealth of literature on calculating risks in museum collections and this "fine tuning" of risk values is covered comprehensively by Robert Waller (see Further Reading).

Drilling down in detail

A risk assessment or the self-assessments above will give you an overview of risks but you may want to know how many herbarium cabinets, insect drawers or spirit jars are not only at risk from agents of deterioration, but how accessible they are and, importantly, what is their value. Exhaustive methodologies for assessing the "health" of natural science collections that have been available since the early 1990s, but while they provide much valuable information, they are time and resource heavy. There was a need for relatively simple assessments that could give a "profile" of the collection on a simple scale based on collections units such as drawers, herbarium cabinets, shelves of jars, etc. The system devised by McGinley (1993) used a simple 10-point scale roughly following the process of improvement through time from a box or jar of unprocessed field material to a fully labelled, digitised and published drawer of insects. This provided a simple graphical representation of how many drawers were currently at levels 1, 2, 3, 4, etc. With this data, targets could be set for improvements, e.g. all drawers will be at minimum of level 4 within the next two years. The method has been modified and repurposed for a variety of collections. The basic principles of ease of gathering data and providing a graphic impression of the current and desired state of a collection has been employed to good purpose by the Smithsonian National Museum of Natural History.

In the mid 2000s, the Smithsonian National Museum of Natural History developed a collections assessment methodology originally called by the informal name "move the dots". It is now simply called the Collections Assessment. This schema was developed to answer basic questions about the appropriateness of the collection relative to the museum's mission, and the readiness of the collection for use. Work began with clarifying the intellectual framework for the entire collection, and continued with an exercise to group the collection for assessment purposes (see grouping discussion on page 49). Assessment criteria were established to indicate physical condition and informational status. These two domains of collection status were complemented by criteria that allow the ranking of the importance of a collection. The museum uses rankings of 1–5 with a score of 5 being optimal and 1 being absent or poor. For each criterion, a rank of 3 indicates an acceptable state. This basic system allows measures of physical condition, informational status or importance to be graphed so that priorities can be quickly identified. Each collection is represented by a dot on a graph. This system is now used on an annual basis. One of its immediate benefits is the transparent and easy way that results and change over time can be communicated graphically by a shift in dots from the left of the graph to the right (See Figure 3.1).

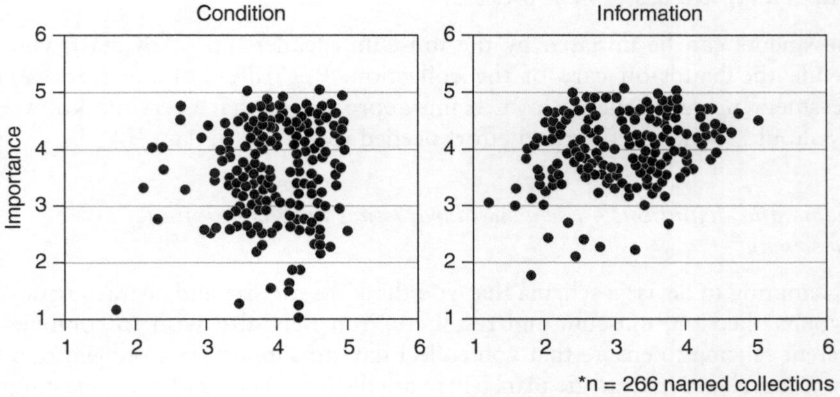

Figure 3.1 Example of one of the graphic representations of results of the initial collections assessment survey carried out at the Smithsonian National Museum of Natural History. Each dot represents one of 266 collections units surveyed (e.g. small whales, echinoderm types, frozen mammal tissues) and is scored on its condition and its importance

This assessment methodology has been adopted and used at three other large natural history museums in Europe (Natural History Museum London, Naturalis Leiden and Museum für Naturkunde Berlin). Other Smithsonian museums use a less-detailed version of this schema. The Smithsonian has also developed a schema for assessing the quality of buildings and spaces for collections storage purposes. While not published so far, these schemas have been discussed at conferences and information on them is available by contacting the Smithsonian Institution.

Peer-review assessments

Assessments and reviews can be carried out on collections by external bodies. These are often triggered by an institution's application for recognition as performing above a sector standard. Museum accreditation schemes, such as that used by the Alliance for American Museums or the Associated Museum Assessment Program and Collections Assessment Program, are well established in the United States (see www.aam-us.org/programs/accreditation-excellence-programs/accreditation/). These assessments are thoroughly validated and have been very effective at guiding museums at making improvements. They also link closely to funding. In the UK, the Arts Council England accreditation scheme "sets out nationally-agreed standards, which inspire the confidence of the public and funding and governing bodies" (www.artscouncil.org.uk/supporting-arts-museums-and-libraries/uk-museum-accreditation-scheme).

What is a typical assessment process?

Assessments can be initiated by the museum's leadership or by staff who provide the hands-on care of the collection. Regardless of the start, an assessment project benefits from being approved so that everyone knows they should spend the time and effort needed to get useful data.

Schema and definitions – use a standard: issues with creation of new schema

It is tempting to devise a schema that you think fits the size and characteristics of your collection, museum and resources. You may also wish to combine different schema to ensure that you collect data to support development of a very comprehensive strategic plan. There are distinct benefits to using existing schema, however. You can contact the developers of the schema or other organisations that have used it for advice. You can refer other staff members to publications and conferences where the schema has been discussed and learn about how effective the schema has been when developing plans and achieving results. External reviewers or funders are more likely to be familiar with existing schema. Making a new schema can confirm internal biases – if the assessment is conducted only once, you may not detect your internal bias. If you conduct your assessment every year, you may still have bias but at least it is in the same direction – you can still identify improvements over time. If your concern is that different reviewers will score the same phenomenon in different ways, you can reduce the likelihood of that problem by training the reviewers, conducting a pilot assessment and using written standards. You could also use a cross-museum team, for example one person from each collections group, to conduct a museum-wide assessment. In smaller institutions where there may only be one or two staff members there is value in forming consortia with other small institutions and sharing the assessment load between the group.

What makes a good assessment?

Whether you opt for a broad-brush survey or a more detailed approach; whether you devise your own or use an existing schema, it is important that they are:

- transparent – data and methods are accessed by all relevant parties and are consistent;
- sustainable – you have the resources to carry them out and repeat if necessary;
- repeatable – you can easily resurvey your collection and measure changes over time;
- computable – you can analyse them easily without excessive manual interpretation.

It is easier to meet these criteria if collecting and maintaining data is supported and integrated with institutional systems. You can categorise your data to match different types of strategic goals. For example internal collections (e.g. conservation, space planning, digitisation, research and access), internal performance (e.g. staff competencies and museum governance) and external policy (e.g. accreditation, policy and wider research/data needs).

Whichever schema you adopt, once you have collected data it needs to be interpreted. This can take the form of reviewing the difference between the newly observed current state and the desired state, reviewing the results in graphical form or using any formulas included in the schema.

Managing forests not trees – selecting units of analysis

When assessing a collection there can be a tendency to focus on the object or specimen (item) level and gather very detailed data. This can come into conflict with the need for a general or high level of information that is accurate, but not necessarily precise. You can think of this as managing and collecting data on a whole forest, not each tree in that forest. This approach has advantages and detractions. Aggregating collections for analytical purposes is typically faster than the more detailed act of examining each item and it is easier to communicate with stakeholders using aggregate data. Conversely, staff sometimes resist aggregating data because they see specific differences or qualities that they believe are significant for analytical rigour. They may also wish to focus on a more detailed approach because they enjoy examining the collections and gathering information, and demonstrating their expertise. It can also be difficult to agree on how to aggregate data, especially how large or small a grouping is appropriate. Exceptions to the aggregation approach would be collecting data on a small, exceptionally rare, valuable (scientifically, monetarily or iconic) collection. While every collection has value, most can be assessed in the aggregate or "forest" approach rather than the individual or "tree" approach.

For most scientific collections, use of an aggregation approach means the collection needs to be grouped according to some organising principles. Those might be taxonomic, geographic, cultural affiliation, time period, by collector or collecting event or by specialized preservation need – and each of these criteria will influence how you can interpret the resulting data. Art and history collections can be grouped too, according to organising criteria that makes sense within those disciplines.

An important factor to consider is whether the grouping is so large that change cannot be identified within each review cycle. If the group is comprised of hundreds of thousands or millions of specimens, such as insects, it can be difficult to see incremental change if the group is simply "insects". Generally speaking, it is easier to demonstrate incremental progress if the collection is grouped into smaller categories. You could think of this as grouping all the oaks or beeches or firs in the forest, instead of each individual tree or the forest in its entirety – or, to follow the insect example, you could group an insect collection by class, order, family, or genus. If your assessment and resulting strategy is set at the level of all specimens from a continent, instead of country or province, it may be difficult to demonstrate progress in your strategy to make them accessible via digitisation. An extreme example is assessing the condition of an elephant specimen (one item) and monitoring conservation procedures and achievement versus assessing all the butterflies in a collection (50 drawers containing between 5–40 butterflies each) and monitoring the conservation procedures to rehouse and repair those insects. The ease of monitoring improvement activities should be balanced against the effort to assess many groups containing a small number of objects or specimens. As a practical matter, the staff managing the collection and leading the strategy should have a strong role in determining how the collection is grouped when planning the assessment.

The tyranny of numbers

Collections managers are often asked to provide data to senior managers and funding bodies on how many specimens there are in the collection. It can be a controversial topic and a point of disagreement. Some may be concerned with consistency of counts with prior years' reports. Others might have an interest in inflating figures to impress or influence opinions on how dire a problem is or how successful work has been. Some may not need precision, but instead will accept numbers at orders of magnitude. And there are those looking for realistic numbers to help with assessment and planning. Interestingly some international data collations record numbers of specimens in log base ten groupings: 1 = 10–99, 2 = 100–999, etc. Use of a log scale in itself has pros and cons. It is easy to quickly identify the number of items in a collection, so that different collections can be compared quickly. Large collections can be characterised quickly using such scales. However, use of log scales will not support precise counts and it makes arithmetic analyses difficult to justify.

Exercise

How do you typically count the collection? Do you have an official count that you must report on a regular basis or do you use different methods for different purposes? List the ways you count now and what current uses you make of that data. What other methods would be required, if any, to strategically plan for an improvement project?

When considering counting methods and strategic planning, it can be very important to select a reproducible and sustainable method. Plan ahead so that subsequent assessments or other studies can use the data, or you can gather data on a regular basis so that you can demonstrate change through time. Data should be gathered and reported using the same principles each year, although tools may change.

Museums have one-to-one, one-to-many, many-to-one and many-to-many relationships in collections and specimen parts and their various record types, thus frequently there is no single, consistent one-to-one relationship across the collections between item, descriptive record and surrogate (image). Some items are part of a larger group or "lot" that is considered to be a coherent set, e.g. six shrimp of the same species in a jar from the same locality, teacups and saucers from a tea set or slide-mounted thin sections from one worm. Some items have related items that can be counted and managed as unique entities, e.g. the skin of a squirrel and the frozen sample of that squirrel's blood. Museums having smaller collections likely find it easier to be more accurate when counting their collections, especially if their annual acquisition rate is low. Once an accurate baseline count is established, the annual increase and decrease is more easily managed. The challenge of counting is usually greater as the collection size increases. But smaller collections can still have the complexity of many-to-many relationships and still have the challenge of explaining what different numbers mean to various stakeholders.

One large museum, the Smithsonian National Natural History Museum, has counted its collection using both precision and estimation. Traditionally, the museum has accepted some rounding error or approximation in the collections counts because, with some specific exceptions such as high value items like gems, those counts were used to track trends. How you count may be affected by your collections information management system. Does it use parent-child records? Or does each physical item have its own record? In any regard, it may be helpful to have an explanation for the different ways you count, and the relationship between specimens and catalogue or digital records. If you make estimations or use sampling strategies, it is wise to document how those estimates are made for explanatory as well as future uses. You may wish to record your museum's standards and the typical counts for the following:

- number of specimens to number of lots – for estimation purposes, this is typically X specimens per lot;
- number of associated items, e.g. a work of art and its frame, multiple prints or other physical items;
- number of associated surrogates, e.g. images;
- number of intended records to number of extant records, e.g. museum Z expects that about n thousand records will appropriately represent the X million items in the collection.

Certainly, if you are able to have the precision of counting or assessing every item, that is a great opportunity for gathering very precise information. If you are not reasonably able to count or assess each item, you can still obtain highly usable results.

What makes a good strategic plan?

The purpose of a strategic plan and the assessment method used to gather information will have an effect on the way you develop a project. The unique situation of each museum will play a role in assessment, strategic planning and management. The process outlined below is intended to give you a general starting point and help you consider some of the questions and topics that typically arise.

Clarify purpose of institution and intellectual framework for collection

Whenever you begin a strategic plan or a project that uses significant resources (time, money, space, etc.), it can be very helpful to revisit the purpose of your museum and your collection, write down how the planned work is an expression of purpose and why it is important to do the work now. If you are developing a strategic acquisition plan, you might ask questions such as:

- Do we have the collections that fit our mission?
- Will we need to continue collecting new specimens for a time series analysis to:
  - reflect biodiversity from a region that is poorly represented in the collection or from new locations, now that the technology is affordable?
  - document the rise and modification of a diagnostic motif in artworks?
  - hold an example of a new technique of manufacture?

If you are planning a preventive care improvement project, you might ask, "Do we have risks due to failing building enclosure and environment that negatively impact the preservation of collection items? Do we have cabinets or shelves that provide adequate support and enclosures?" If you are gathering data to support strategic planning for a digitisation project, you might ask, "Who typically uses our data and how do they get access to it?" Regardless of the strategic action you wish to take, make sure you can communicate clearly and consistently to others about how that work fulfils the museum's mission and why it is important to address now.

Validate the desired state and goal you are pursuing

This step may focus on a very specific aspect of collections management, such as rehousing of fluid preserved specimens, or it may encompass a broad range of preventive care, digital access, risk management, enhanced usage and so on. The important point is to be able to clearly and consistently communicate the desired state or goal in measurable terms.

What do you want to do with the assessment data? Is there a particular problem you intend to solve? What are your fundamental questions? Responses to these questions will affect what kind of assessment is appropriate for use or if you need to use multiple schemas to gather the information you need. Then select the appropriate assessment schema.

Organise your project

1 After you have identified the topic of your strategic plan and basic goals, consult any stakeholders to gain their insights, consider their needs and requirements and include them in any resource and schedule impacts that occur as a result of the plan's action steps.
2 Establish your team. What staff will be involved? Who will fill what roles? Are there experts who should lead some elements? Are there staff that need an opportunity to develop new skills and confidence through a challenging assignment?
3 Set your schedule. If you identify when an action needs to be completed, you can work backwards to set start and incremental progress dates. This will help you see if your schedule is realistic. Do you have a requirement to demonstrate strategic improvements by a specific date? It is helpful to work with your team and your stakeholders to set the schedule so that everyone understands how the project will affect other work and resources.
4 Establish your data collection method. This will be related to the kind of assessment you conduct, but regardless of method you will need to know who is collecting data, using what media and in what format. Will staff use a spreadsheet, data base or data entry form? Or pen and paper? What are the risks of different media and formats? Will the data be incorporated into a museum system or kept in a standalone file? As you work on this step, it is helpful to discuss risks such as inaccurate data, non-standard data that cannot be interpreted or analysed, or data loss so that you have mitigation strategies in place.
5 Establish a place to maintain data. Information that results from assessments can serve a wide variety of purposes, but only if the data are reliably accessible and maintained. Consider who controls the preservation and access to the data. For large projects that use significant resources, it is advisable that data be maintained in museum systems.
6 Reports. What format will you use to share assessment data? Typically, stakeholders and decision makers will ask to see the findings of the assessment. They will appreciate seeing data in a form that can be quickly understood rather than a long listing of raw data. Thinking ahead to how the data can be displayed in graphical form will pay

benefits when it is time to write the recommendations and track progress. As you assemble the project team, consider this aspect and engage colleagues who can bring information management expertise to the team.

Gather data

The data-gathering step in an assessment is a double-edged sword. It serves as a wonderful opportunity for building a shared understanding of the collection and museum standards. Depending on the assessment schema, vast quantities of valuable information can be gained. The teamwork can be fun, too. But gathering data should not take so much time and focus that everyone loses sight of the purpose, which is to gather and interpret information, compare it to the desired state, develop recommendations and action steps, and achieve the strategic goal. If you are using an assessment schema for the first time, or integrating new team members into an existing assessment process, it is helpful to conduct a short pilot project to test how the assessment is conducted and validate a "first cut" of the data. You can refine the process; clarify expectations regarding the schedule and time allotted for assessment. You can review early results with the team and ask critical questions as a way of ensuring that everyone is using the assessment criteria the same way. So, while you will want to keep the project moving towards the goal, it is very important to ensure that you are collecting accurate data before you need to interpret it.

Vet, interpret and visualise your data

Once all your assessment data is collected, it is worthwhile to review it with the project team, colleagues and stakeholders. They will appreciate knowing that progress is being made, and you can use this opportunity to ask for clarifications about any patterns or working interpretations that are emerging. This review step is also helpful as a final point for checking that assessment criteria have been used properly.

The assessment schema you use may include a formula or other means of analysing the data for interpretive purposes. In its most simple form, you can compare the assessment data to the desired state of the collection for that topic, criterion or problem.

Frequently, assessment data are gathered as numeric scores on a chart or spreadsheet. Sometimes scores are made in tick boxes on a ranked listing. Other assessments use short narrative responses. Regardless of how the data are gathered, it is best to develop some means of seeing the data in graphical form as a part of beginning the interpretation process. Many computer applications include a function that supports creation of graphs, which is a good place to start. Some museums have experts on staff that can help with interpreting data through statistical analyses or visualisations. If you engage experts in the planning process and selection of the assessment schema, you will speed up the process of getting visualisations that are useful.

Plan ahead and consider what kinds of visualisations you will need to inform others of the assessment results. As part of your pilot project and data vetting discussions, you can ask the stakeholders about their preferences for reviewing data. Are there formats that are easier to understand? Are there formats that are easier to reproduce in presentations and publications?

Developing recommendations

As you review the results of your assessment, what are the areas where there is a clear difference between the current and desired state of the collection? These can form the natural starting point for developing recommendations for actions to take to achieve your goal. Different stakeholders and experts within the museum will have different potential recommendations, so it is beneficial to include reviews of the assessment with them. Typical recommendations when assessing condition will focus on preventive conservation means that can be taken, especially measures that solve a problem once, not just implement a procedure. Example recommendations for a preventive conservation assessment could be:

- improve the environmental controls in all collections store rooms having a score that indicates unstable conditions;
- replace all cabinets that have warped doors;
- relocate all collections away from exterior walls.

Many museums publish their strategic plans. Reading these plans and discussing them with colleagues is an excellent step in preparing for your own strategic planning process. It can help you and less experienced teammates gain understanding and confidence in developing recommendations and their associated action plans.

Develop an action plan

Action plans are a series of specific steps that will be taken to accomplish the recommendations. Action plans state what work will be done, by whom, according to what schedule. They can include a statement about the resources that will be available to accomplish the work. Action plans should be measurable. It should be clear whether the work in each step has been accomplished, making action plans an essential element for tracking the progress of your strategic plan.

Following the example recommendations above, sample action steps for the second recommendation are outlined below. Clearly, they are not exhaustive but they do illustrate the relationship between the recommendation and action.

Recommendation: Replace all cabinets that have warped doors.

Action 1: Using the sampling data in the assessment, review each cabinet in the collections where warped doors were observed. Record the count of failing cabinets and their location and general contents on an electronic data form. Review must be completed in two weeks. Task leader is Ms. C.

Action 2: After completing the review where warped cabinets were observed, conduct a review of the collections where warped cabinets were not observed in the initial assessment. Record information using the same electronic form as used previously. Review must be completed in four weeks. Task leader is Ms. C.

Action 3: Using the detailed listing of failing cabinets, prepare a one page overview of the relative risks for the collections, taking into account the agents of deterioration most prevalent in the museum. The written and graphical risks overview should be ready for review within two weeks of completing Action steps 1 and 2.

Action 4: The management team will review the risks for each collection needing replacement cabinets and will develop a prioritisation for purchasing new cabinets. The prioritisation will be reviewed and agreed within two weeks of the risks overview presentation. Task leader is Ms. B, stakeholders are the collections management team and the finance officer.

Action 5: The finance team will order the cabinets from an agreed manufacturer within five days of receiving the prioritised listing. Task leader is Mr. M.

We have included an example of an action plan for a legal documentation project in Appendix 3.3.

How can I make the planning process easier?

Using assessments to assist with strategic planning can seem daunting, especially if the collection is large, or the number of staff who can participate in the assessment activity is small. Here, we provide some suggestions for how you may ease the process.

Can you group collections to make planning simpler?

Natural science collections can be astonishingly diverse and reflect the scholarly history of many disparate disciplines. The size and diversity of the collections, coupled with the intent to assess and plan on a collection rather than specimen level, means that clustering or grouping specimens into defined pragmatic or nominal subunit collections is essential. For biological collections in particular, the taxonomic hierarchy offers many ways to subdivide collections. Among entomologists, use of arthropod orders are a natural smaller choice. Vertebrate collections are traditionally subdivided into fishes, mammals, birds, with reptiles and amphibians together. Invertebrate zoologists often speak of phyla.

Example: Grouping collections by management units

When conducting its assessment, the Smithsonian National Museum of Natural History departments were encouraged to group collections in a way that was consistent with how they are managed. They could also single out collections historically recognised as important, for example fossil and extant marine mammals, gems, mosquitoes or their coral and crustacean collections. These are examples of collections for which the Smithsonian National Museum of Natural History is particularly noted and whose condition and improvement museum management should particularly track. Other criteria that were frequently used when grouping included time period, culture, taxonomy and locality. In any event, the resulting list of collections encompasses all specimens, objects and other collection materials at the Smithsonian National Museum of Natural History and satellite facilities.

RACI approach to communication

The assessment process, indeed every aspect of strategic planning and management, can benefit from planning how communication will occur and how information from others will be solicited. We all have the challenge to ensure that work is done, that the appropriate people have contributed to information gathering and decision making and that everyone is aware of the topic.

A management concept known as RACI (Morgan, 2008) can help project leaders and managers ensure that everyone is clear on their role. The RACI approach defines who is:

- responsible – the person or group doing the work;
- accountable – the single person who will get credit or blame for the status of the work;
- consulted – people whose opinions or input will influence the project or work;
- informed – people who will receive information about the status of the project or work.

Have you ever encountered a situation when someone was unhappy that no one requested their opinion? Or a situation where an opinion was requested but never provided? Or where the team leader doing the work thought she had to carry the whole burden alone? Instead of facing these kinds of questions or difficulties, a team leader or manager could use the RACI approach so that everyone associated with the project understands their role. Using the cabinetry replacement recommendation above, here is an example of the RACI approach:

For Action 1:

- Responsible – Ms. C and technicians A, B, and C.
- Accountable – collections management leader Ms. B.
- Consulted – Ms. C will set the definition of "warped" for the project.
- Informed – everyone in the collections department will be informed that a detailed review is being performed.

For Action 2:

- Responsible – Ms. C and technicians A, B, and C.
- Accountable – collections management leader Ms. B.
- Consulted – Ms. C will set the definition of "warped" for the project.
- Informed – everyone in the collections department will be informed that a detailed review is being performed.

For Action 3:

- Responsible – Ms. C and technicians A, B, and C.
- Accountable – collections management leader Ms. B.
- Consulted – conservator Dr. H will train Ms. C and the technicians and set the standards regarding agents of deterioration. Subject matter experts for each collection will be consulted for their input on the agents most prevalent and damaging in their collection. This information will be provided to Ms. C. The collections managers team will be consulted for their preferences regarding format for the overview of relative risks.
- Informed – everyone in the collections department will be informed that a detailed review is being performed. After collections management leader Ms. B accepts the overview, it will be provided to everyone in the collections department as part of a presentation to keep them informed of the progress of the strategic plan.

For Action 4:

- Responsible – Ms. B and finance officer W.
- Accountable – Ms. B.
- Consulted – conservator Dr. H and subject experts for each collection regarding risks of delay per prevalent agents of deterioration. Finance leader M will be consulted for budgetary impacts

and options regarding purchasing all cabinets at once, or in phases based upon prioritised risks.

- Informed – everyone in the collections department will be informed that a detailed review is being performed. After collections management leader Ms. B accepts the overview, it will be provided to everyone in the collections department as part of a presentation to keep them informed of the progress of the strategic plan.

For Action 5:

- Responsible – the purchasing team.
- Accountable – finance officer W.
- Consulted – conservator Dr. H regarding appropriate cabinet vendors.
- Informed – everyone in the collections department.

Working with stakeholders

Your stakeholders are the people who have a vested interest in the collections. They can be staff working to provide care and access, users of the collections and people with administrative or fiduciary responsibility for the collections among other types of stakeholders. Are you getting the input you need from your stakeholders? If not, consider who you are asking. Are you being clear that you are consulting someone for their input, and that input will be important in decision making?

- How specific are your questions?
- Are you asking closed "yes or no" questions or open questions that lead to a detailed response?
- Are you asking more junior staff or volunteers involved in the project to provide input?
- Do stakeholders feel they can provide critical or constructive comments or are they avoiding difficult conversations?

Regular communication with stakeholders will help you to keep a broad perspective on your topic so you can avoid reinforcing internal biases. Getting input from external users of the collections, or external review committees or advisors can be very helpful. They can help you examine your assumptions so that you avoid uncritical thinking.

How can my team think and act strategically?

Before embarking on gathering data and developing a strategy, it is worth thinking about the people who are establishing the concept, setting up assessments and identifying the need to improve and then implementing and monitoring progress. In a very small collection organisation these could be one or two people, in others individuals from across an organisation brought together for these tasks, or dedicated teams.

Do you think that you have all the skills and behaviours or competencies to achieve all this? Firstly, do all your staff, but particularly your decision makers, think and act at the "museum level"? It is often the case, especially in larger museums, that the staff brought into a collections planning strategy might be highly focussed on the collection which is their responsibility. Knowing what works for moths is fine but could this equally apply across the collection? In our experience, the earlier these issues are addressed the better. In Chapter 5 we will look at how competency frameworks can help you to know the skills and knowledge of not only your team but others across your organisation and use this to select the most appropriate people for the task.

Exercise

How can you discover whether your team's decisions, but in fact your decisions as well, are fact-based or personality driven? How do your team members prioritise their work? Can you indicate whether it is based on personal interests or an overall strategy supported by policies and procedures? Looking across the team, how many are equipped with information they need for problem solving? Give examples.

Can those setting and implementing the strategy communicate using consistent terminology? Communication is essential but it can be hampered by different terminologies that have grown up in silo-like departments over many years.

Strongly following on from this, how would the strategic process continue if key staff were unavailable for any reason – how would you make informed decisions? This is often referred to as the "bus test". Crudely put, if a staff member was run over by a bus could the process continue or would key information be lost and not be obtainable anywhere else within a reasonable time frame. If the answer to any of the above is "no" then how can you move ahead with planning (see Chapter 5)?

Systems and processes, not heroism

Do you find that your collections work routinely requires marathon-like sessions that greatly exceed the normal work day? Do these sessions feel like everyone is responding to an emergency and do people crave the recognition of being a hero, or want to be "magicians" by finding last minute solutions?

Do your days feel chaotic and overwhelming? If so, you may be suffering from manufactured emergencies and work practices based on heroism, not systems, processes and plans. True emergencies do arise and these certainly demand special attention but they should be rare. As far as possible, the predictable, scheduled work of the museum should not require regular heroic action because of poor communication, planning and scheduling. Managers at any level or organisation size can avoid this by understanding the work to be done and developing realistic plans that take available resources and information about the task at hand into consideration. Realistic planning usually results in reproducible, predictable actions that can be budgeted and scheduled. When the museum makes a habit of developing plans and using reproducible processes, systems emerge and work feels less chaotic and overwhelming. Use of systems allows everyone to be accountable for work and use of resources. Systems also support development of a predictable workload for staff. These systems can be on any scale and are not exclusive to big organisations.

How can I prioritise actions in my plan?

Your assessment may reveal multiple problems of similar importance or magnitude. Weighing options, balancing competing demands and ultimately setting priorities is the crux of strategic decision making for collections. Every natural science collection has challenges, and every organisation has desired improvements. We all have some "givens": collections are always in a slow or rapid decline towards decay and loss. If they are frequently used, collections are always tending towards disorder or chaos. Increasingly, users want more information about specimens. Costs tend to rise and resources tend to become scarcer. Sometimes the challenge comes from competing needs between different kinds of users, such as researchers and educators, or exhibition developers and conservators.

What do you need to do before making a decision on priorities? Here we list some of the points to take into consideration:

- What is the purpose of your collection?
- Is there a law or policy that can guide priorities?
- What is the potential for competing needs among your stakeholders that could influence your priorities?
- What data do you have about the collection's status?
- What is the availability of existing resources?
- What are the risks of inaction – can you recover if things go really badly?
- Are there larger political, historical or scientific considerations?
- Can you consult your stakeholders with enough time to make an informed decision?
- Do you have the support you need from museum leadership?

Exercise

How do you prioritise decisions now? Who decides? How frequently are decisions reviewed? Would everyone on your team prioritise a project the same way? Would they agree on the points to consider when setting priorities? Would you weight the factors listed above and if so how?

How to decide between spending scarce resources on similarly needy collections?

This is a common dilemma faced by managers in all organisations large and small. There is no obvious answer as every situation is different but having as much data available on the collections beyond their immediate needs will help. We hope this example will assist you in these challenging situations.

Example: Deciding which collection to resource

Collections A, B and C all needed new cabinets. All contained rare and unique specimens of high scientific and historic importance. All were relevant and regularly used by researchers. All were subject to the same agents of deterioration and all had equivalent scores on the recent assessment. Collection A needed 25 cabinets, collection B needed 48 cabinets and collection C needed 15, but the funds available could only cover 66 cabinets or 75% of the total cost. The collections leader had the following choices:

- Option 1: buy all the cabinets for collection A and B, and defer the cabinets for C until the next year.
- Option 2: buy as many new cabinets as the budget would allow, and distribute them equally across all three subgroups.
- Option 3: wait one year for additional funding and then buy all cabinets for collections A, B and C.

What other factors could contribute to decision making? The collections leader knew that collection A was going to be moved to a new location in two years. She also knew that a large sum had been spent on collection B within the last three years. Collection C's staff were very busy with a large digitisation project and were unlikely to be able to drop that work to move collection C into new cabinets. Collection A was not duplicated in any other museum in the country, and along with the planned move, a grant proposal was being reviewed that might result in external funds for preventive conservation. After consulting with stakeholders, the collections leader decided to spend the funds on collections B and C (a total of 63 cabinets) and slowly begin the purchase of cabinets for collection A. What would you do?

Assessments of value for prioritisation purposes (not valuation)

We all make judgments all the time, sometimes consciously, sometimes unconsciously. We know what we think is beautiful, amusing, smart, worthwhile or valuable. The idea that all collections are valuable is true to some extent, but some are more important than others. This dimension of "importance" or "value" can be a critical factor that affects prioritisation. How can you use an objective means to decide upon value? We recommend you solicit input from the stakeholders at your museum. Revisit the purpose of the museum. Consider whether the importance of a specific collection can change through time. Are there emerging technologies that will reveal new data from existing collections? What external factors could influence the perception of importance? See Chapter 2 for references and further discussion of value.

Exercise

Create a matrix or decision tree for the dimensions of "importance" or "value" at your museum. Would different stakeholders assign measures of value the same way? Are there any measures of importance that are established by laws and policies?

Prioritisation and management

Strategic planning and management always has an element of monitoring work and confirming that decisions and actions stay true to plans. You can see the importance of monitoring work when looking at the use of resources. Resources exist to achieve the purpose of your museum. Allocation of resources (funds, space, time) should be consistent with purposes and needs of the organisation and any governance requirements. Strategic plans should indicate what resources are required for execution, the impacts of partial funding, and key performance indicators related to resources beyond stating what resources were expended over a period of time. Expending budget should be rebalanced periodically if the aims of the organisation are not being met. For many collections professionals the resource that is most within their control is time. See the example below for how a decision about time was ultimately not a good one.

Example: Resources are more than just funding

Five staff were assigned to assist a researcher in answering an information question for a governmental agency. Those five staff worked for three months to gather existing analogue data and present it in digital form. Time expended equalled 15 months (5*3). At a cost of $ 3,000 per month, the labour costs for those staff alone on the project was $ 45,000, not including the labour costs of the researcher, the information managers and any other staff involved in answering the basic question for the government agency. After hearing about the allocation of staff time, the programme manager asked questions about the benefits of the project. What were the other uses of the resulting information? There were no uses other than answering the request and having that information available for future but undefined purposes. No new collections records were created and there was no scientific publication. What was the result of providing data to the government agency? The museum received a response of "thank you". There was no direct increase or impact on public understanding, no new allocations of resource, no new scientific understanding. In essence, it was a very expensive use of resources for little tangible outcome, making it not an effective use of resources. The supervisor of the five staff and the programme manager agreed that in the future, staff time costs would be calculated prior to assigning staff to a project.

Although this example may be a salutary lesson, strategic use of resources is about more than just financial resources. Staff time and space can be more precious than funds. Undirected staff who decide to spend their work time on tasks and projects that aren't strategic are misusing the museum's resources. Using climate controlled and highly secure collections storage space for unmanaged office equipment, supplies, broken chairs, etc. is an inefficient use of that resource and prevents it from being used appropriately. In the absence of a strategic plan, people will do what they think is most appropriate or enjoyable. If the staff are well aligned with the purpose of the museum, this may not be a critical problem. If the staff are not aligned with the museum's leadership or its purposes and priorities, museum resources can be used inefficiently at best or at worst be used in a fraudulent way. When resources are misused, collections professionals can return to their strategic plan and its action steps to resume orderly, planned work.

Are there templates that I can use to organise my actions?

One way to organise and prioritise preservation actions is to group those actions according to a hierarchical schema. A preventive preservation hierarchy can lead to a preservation or physical care plan that is part of the museum's collections plan, complementing other plans such as digitisation plans and various building renovation and collections relocation plans. The preservation schema can draw on information gathered through collections assessments. We have included an example of a hierarchical schema for a preservation improvement project in Appendix 3.2.

How can I know that I am meeting my collections objectives in the longer term?

We looked earlier at a range of assessments that can be used to show the current state of the collections and to provide objective data to support strategy development and prioritisation. However, how can you detect what effect there has been in a collection's or institution's performance on a year-by-year basis in response to a strategic action? Alternatively, can you access information on day-to-day business, e.g. are visitors increasing and are you delivering the services and internal needs that are expected? Progress (or lack of it) can be measured by collecting specific data on products and activities that are essential to achieving goals. For example, a museum might ask if the numbers of users have increased after staff have implemented an advocacy programme. Has digital access affected the number of visitors and loans?

Performance Indicators (PIs) are a familiar and much used business methodology for measuring change. Since the 1980s, museums and other public sector institutions, where performance cannot be measured simply by profit, have developed relevant PIs for use by their managers, funding bodies and regulators to demonstrate accountability, efficiency and effectiveness. Collections managers often use PIs in this context, rather than to inform policy and planning in collections management, for which these traditional measures are often of limited value.

A commonly used performance indicator in museums is the number of loans per year. This is often demanded by funding bodies and senior managers. But what is it an indicator of? Did you have the objective to send out x number of specimens to x number of institutions in the last year? If so how could you change it? Maybe loans are driven by external factors beyond your control. Number of loans is an indicator of activity and may help in allocating staff time in a future planning cycle. Perhaps a more useful indicator of performance would be how many loans have been processed and sent within one week? This measures how efficient the loan system is and could be a target to make the museum more accessible and professional in its dealing with requests from external users.

Exercise

Think about some of your key activities in management, preservation, access and promotion of your collections. These could include registration of acquisitions, recovery of overdue loans, number of specimens rehoused or jars of fluid preserved specimens that are refilled, number of new blog posts about collections or other activities.

- What indicators do you commonly use or would consider using?
- How well do they tell you how you have performed against the objectives?
- What might be more useful to you for that purpose?

When you have drawn up your list of PIs, look at them in the light of the list of requirements of a good indicator below. Does your list meet the requirements?

You may have found that while some of these metrics indicate change they do not necessarily indicate changes that you have much control over.

The SYNTHESYS project (Guiraud & Gregson, 2013), devised a new set of transferable collections management PIs and definitions that could be used by many different museums. They were designed to allow institutions to create their own subset of indicators for internal management, which will prove useful in developing new, more efficient ways of working. They were developed in 2012 with input from museums and science sector organisations to seek new methods to improve workflows and processes to maximise the efficient use of staff time and other resources.

How can performance indicators help?

Areas where performance measurement contributes to museum management activities include:

- formation and implementation of policy;
- planning and budgeting of service provision;
- service standards;
- efficient use of resources;
- control and influence over decision making.

Collections managers can also use PIs to identify how performance has changed over time, compare departments within an institution or between similar museums.

The process of determining which areas of collections management should be monitored, and setting standards, can actually encourage discussion of what collections management should be in a museum and what qualities all museums in the same sector (e.g. natural history) should have. Common PIs have the potential to unite museums in a sector in a common goal. Also within an institution, developing targets and the ability to measure performance helps clarify the mission and sense of purpose of collections managers, providing evidence of their commitment to excellence and improving their management systems.

What makes a good performance indicator?

- Consistency and comparability – definitions consistent over time and between institutions to enable comparison of like with like.
- Clarity – simple, well-defined, easily understood.
- Controllability – only aspects of performance over which there is control should be measured.
- Limited – concentrate on a limited number of PIs that give the most valuable collections management information.
- Feasibility – can the PIs be measured easily?

What risks are associated with using performance indicators?

The SYNTHESYS project identified several risks associated with indicators that collections managers should keep in mind when developing them:

Measurement

- PIs from different institutions or departments must be comparable – definitions of units of measure and rules must be exactly the same.

Behavioural

- PIs must be kept in perspective to prevent the focus of collections management switching from actual mission accomplishment to the PI system itself.
- PIs must help enough to be worth the effort.
- PIs should not focus too much on managerial control or appraisal aspects, or performance measurement could be regarded by staff with apprehension or hostility.

Conceptual

Many of the qualities of good museums cannot be measured numerically, for example PIs cannot easily measure importance or quality and should not be read as more than a partial portrait. At best, they can only identify characteristics of efficiency that lend themselves to numerical analysis, and give some indication of whether a collection is well managed.

What strategic questions are others considering?

No matter what the size of the organisation or collection, strategic planning and management can be challenging. Sometimes it is reassuring to know that others are facing the same challenge, even if the scale of the topic is different. It can also be helpful to learn about the strategic topics at other institutions as a way of staying current with the profession, avoiding stale thinking or identifying potential opportunities for collaboration or learning. While in no way exhaustive or universally applicable, here are some sample questions:

- Is it better to have many basic digital catalogue records or a smaller number of enhanced digital catalogue records?
- Should we process the entire cataloguing backlog to our current standard or do a "first pass" process of basic organisation and cataloguing to promote discovery of these resources?
- We have limited space. Should we save it for a new kind of collection or reserve it to use as existing collections increase in numbers?
- Researchers are talking about a new analytical technique. What are the preservation requirements and storage needs for the specimens being collected for use with that technique?
- Is it better to use funds for salaries to hire a few experts and supplement our staffing needs with students, interns, volunteers and contractors to do basic work, or follow our traditional staffing practices?
- If I use a less predictable crowd sourcing approach to meet data transcription needs for my project am I getting community engagement at the cost of timely access to data?
- What is the importance of serendipity in collections access? Is this an old fashioned question in the digital age?
- How can I ensure that I know what new users need from our data?
- How do I promote a deep understanding of the collections that can be transferred to others?

Summary

Collections management as a profession responds to community standards for preservation, access and use. The adept collections professional knows that a strategic approach is necessary given the many, often competing, demands for resources. We hope that through this chapter you see the fundamental relationship between purpose, data, strategic planning and management. Strategy and management involve a continual cycle, starting with purpose and desired changes, through data gathering, analysis, recommendation, prioritisation, actions, monitoring and review – and then beginning anew as opportunities and needs arise. In some museums there are multiple strategic plans underway although one may be the overarching plan. In this way, strategic planning becomes a way of thinking and working, with the museum staff asking, how can this be better? What is the most important thing to do to meet our purpose? What is the strategic action to take right now given our situation?

As a final thought, we encourage you to add the element of "celebration" to the strategic planning cycle. Everyone, regardless of their role, benefits from a public acknowledgement when work is well done. Celebration gives us the opportunity to make note of new skills gained among staff, exemplary performance, problems solved and new opportunities made possible by improved capacities. Including the act of celebration makes strategic planning and management a positive and joyful thing.

References

Collins, C.J., Cornish, L.P., Huxley, R., and Owens, S.J., 2006, SYNTHESYS Network Activity C – Assessing Standards of Collections in European Museums. *Collections Forum*, 21 (1 & 2): 5–18.

Gardner, J.B., and Merritt, E.E., 2004, *The AAM Guide to Collections Planning*. Washington: American Association of Museums.

Guiraud, M., and Gregson, J., 2013, SYNTHESYS Performance Indicators for Natural History Collections Management. https://wissenschaftliche-sammlungen.de/files/9513/7286/0838/SYNTHESYS_Performance_Indicators.pdf [Accessed 30/3/20]

Hillhouse, S., 2012, *Assess and Manage Risk In Collections Care: A Collections Care How To Guide* (A. Dawson and N. Hutcheson, eds.). Norwich: Norfolk Museums and Archaeology Service. http://sharemuseumseast.org.uk/wp-content/uploads/2018/07/How-to-Guides-Assess-and-Manage-Risk-in-Collections-Care.pdf [Accessed 06/11/19]

McGinley, R.J., 1993, Where's the Management In Collections Management? Planning For Improved Care, Greater Use, and Growth of Collections. In: C.L. Rose, S.L. Williams, and J. Gisbert (eds.), *Congreso Mundial Sobre Preservación y Conservación de Colecciones de Historia Natural. Vol 3. Temas de Actualidad, Iniciativas y Direccions Futuras sobre Preservación y Conservación de Colecciones de Historia Natural* (pp. 309–338). Madrid: Dirección General de Bellas Artes y Archivos.

Morgan, R., 2008, *How to Do RACI Charting and Analysis: A Practical Guide*. www.projectsmart.co.uk/how-to-do-raci-charting-and-analysis.php [Accessed 06/11/19]

Further reading

Elkin, L., and Norris, C.A. (eds.), 2019, *Preventive Conservation: Collection Storage*. New York: Society for the Preservation of Natural History Collections; American

Institute for Conservation of Historic and Artistic Works; Smithsonian Institution; The George Washington University Museum Studies Program.
Favret, C. et al., 2007, Profiling Natural History Collections: A Method For Quantitative and Comparative Health Assessment. *Collection Forum*, 22: 53–65.
Hollis, K.A., Smith, D.M., and Spence, C.R., 2010, Conducting Collection Assessments with an Emphasis on Paleontological Collections. *Collection Forum*, 24: 72–79.
Waller, R.R., 1994, Conservation Risk Assessment: A Strategy for Managing Resources for Preventive Conservation. Preprints of the Contributions to the Ottawa Congress, 12–16 September 1994. In: A. Roy and P. Smith (eds.), *Preventive Conservation: Practice, Theory and Research* (pp. 12–16). London: IIC.
Waller, R.R., 2019, Collection Risk Assessment. In: L. Elkin and C.A. Norris (eds.), *Preventive Conservation: Collection Storage* (pp. 59–90). New York: Society for the Preservation of Natural History Collections; American Institute for Conservation of Historic and Artistic Works; Smithsonian Institution; The George Washington University Museum Studies Program.

APPENDIX 3.1: SYNTHESYS COLLECTIONS SELF-ASSESSMENT TOOL

Table 3.3 lists the benchmarks used in the SYNTHESYS Collections Self-Assessment Tool (CSAT). See www.synthesys.info/network-activities/synthesys3-na2/self-assessment.html.

This table of benchmarks was created as part of the SYNTHESYS Research Infrastructures programme project funded through the European Commission under FP7 call "Synthesis of Systematic Resources" and we would kindly ask that this is acknowledged in any re-use, re-purposing or dissemination of these materials. This table reflects the views only of the authors. The Commission cannot be held responsible for any use which may be made of the information contained therein.

Table 3.3 List of benchmarks used in the SYNTHESYS Collections Self-Assessment Tool (CSAT)

Infrastructure – Management

The aim of this section is to identify the institutions management structure, including policies and procedures, legislation and staff training.

	Benchmark	*Explanation*
1	The institution has a written mission statement that makes a commitment to the preservation and management of its collections.	The mission statement articulates the institution's main purpose and highlights its pledge to safeguard its collections. For example "Our mission is: to maintain and develop our collections, and use them to promote the discovery, understanding, responsible use and enjoyment of the natural world. If this is not a separate statement then, at minimum, this information is clearly identified on a public website or in publicity material".
2	The institution has a basic collections management policy.	This policy (or policies) is a statement of what all staff must do regarding the acquisition and movement of collections. For example accessioning; acquiring; appraising; assessing impact; assessing significance; lending and borrowing; removing.
3	The institution has a basic collections care policy.	This policy (or policies) is a statement of what all staff must do regarding maintaining the condition of collections. For example condition checking; conserving; controlling pests; emergency planning; ensuring health and safety; handling/moving; insuring (including indemnity); managing collections care information; maintaining technical viability; monitoring and controlling environment; risk assessing and managing; packing; securing; storing; transporting and business continuity planning.
4	All staff involved in collection management have been issued with current institutional policy documents.	The institution has clear ways of making sure staff are aware of and understand the policy. They are easily available either as a paper copy or on the institution's internal or external web pages.

(*continued*)

Table 3.3 Cont.

Infrastructure – Management		
5	The institution has agreed environmental standards at which its collections should be stored and displayed.	The institution has agreed conditions such as temperature, relative humidity and light at which the collections should be stored. These are available to all staff and ideally are written down.
6	The institution complies with laws and regulations relevant to collections management and care and has a mechanism to ensure staff are kept up to date with changes.	The institution makes sure it is not breaking the law and that all laws relevant to collections are communicated to all relevant staff.
7	Resources and a budget for collections management and collections care have been allocated to ensure all collections can be maintained.	There are sufficient funds allocated to ensure that collections do not deteriorate. For example there is enough money available to purchase appropriate storage materials and alcohol.
8	There is a clearly defined staff structure with defined roles and responsibilities to cover all major areas of collections management.	Staff have clear job objectives and time allocated to collections management and/or care (even if they have other responsibilities such as research, education). All staff know who is responsible for maintenance of a particular collection and who has overall responsibility.
9	There are procedures and documentation in place to ensure that knowledge is transferred and not lost when staff leave.	Before staff leave they pass on as much knowledge as is feasible to new or existing staff who are going to take responsibility for the particular collections. For example overlapping of staff; training/succession planning or providing written documentation.
10	Staff training needs in areas of collections management and collections care are assessed.	Staff training needs in collection management and collection care such as those listed in the help of Questions 2 and 3 are assessed. For example by interview, regular job reviews, managers' observations.
11	Staff have been trained in, and are aware of, current collections management and collections care procedures and best practice.	Staff responsible for collections care and maintenance know of and are trained in currently accepted methods and mechanisms for ensuring conservation and access (including basic preventative conservation and integrated pest management).

Table 3.3 Cont.

Infrastructure – Management		
12	The institution complies with all local, national and any relevant international (for example EU) Health and Safety statutory regulations	
13	The institution provides health and safety working conditions for all staff and other users working with collections.	There are no hazards to staff or visitors in working areas. For example poorly stored chemicals, dangerously stored collections, boxes or other materials that can fall or trip people.
14	The institution has identified all health and safety risks that are posed by their collections.	Those responsible for the collections know of the hazards presented by the collections. For example sharp spines, poisons, fire, hazardous chemicals, toxic substances (for example mercury, arsenic) and radiation.
15	Staff receive regular training and advice on health and safety issues specific to the collections they are working on.	Staff are fully aware of risks associated with the collections they are working with and have been given appropriate information and skills to deal with them.
16	All policies and procedures relating to staff and visitors are in written form, are available to all relevant staff and are updated regularly.	Polices, for example, for loans and acquisition of collections and procedures, such as how specimens are treated for pest infestations, are recorded in either written or electronic form and available to relevant staff.

Infrastructure – buildings

The aim of this section is to identify how the infrastructure (the building and physical storage rooms) is maintained and secured.

	Question	*Explanation*
1	Repairs are made to the collections buildings as a matter of priority to ensure the collections are protected.	The importance of maintaining buildings holding collections is recognised by those responsible for them and actions taken accordingly. For example there are no holes in external roofs, walls, windows; external doors and windows are well fitting; floors are sealed against water coming in from below.

(*continued*)

Table 3.3 Cont.

Infrastructure – buildings		
2	Resources and a budget for the maintenance of collections buildings and associated equipment have been allocated.	Funding and resources are available to keep buildings and associated equipment in a suitable condition. An example of associated equipment is air conditioning systems. If the institution does not own the buildings that the collections are housed in then this question will be "met" if the owner of the building has a budget and resources allocated to maintain the infrastructure.
3	Collections buildings and associated equipment are inspected on a regular basis and potential risks to collections are identified and reports maintained.	Checks are made at intervals agreed by the institution or for equipment as set out by the manufacturer. For example regular inspection of gutters, drainage systems, air conditioning systems. A written record of these checks and their findings is kept.
4	Procedures are in place to ensure that actions are taken to mitigate any risks identified (in Question 3).	The institution takes preventative action if it has identified a risk to the collection. For example if a roof is found to be leaking then collections are moved until it is fixed, gutters are cleared.
5	Trained maintenance staff monitor all buildings and facilities and have sufficient budget to ensure buildings are maintained to an agreed standard and operate to ensure the environment stays within agreed parameters.	Staff with the appropriate skills, a facilities/maintenance buildings department or external contract staff have the funding necessary to keep the buildings and associated equipment in good working condition. For example environmental parameters such as temperature, light and relative humidity are maintained at appropriate levels.
6	Reports on the condition of the building and conservation advice are used in planning improvements or major developments.	Specialist advice and reports are available and are integrated into decision-making processes to ensure any changes or improvements to buildings provide appropriate conditions for storing collections.
7	Access points to the collections building such as doors, lifts, windows and ventilation are secure.	Unauthorised people are prevented from gaining access to the collections by appropriate physical barriers. For example secure doors and windows.

Table 3.3 Cont.

Infrastructure – buildings		
8	Collections are kept in a secure environment, including those that are on display/exhibition.	Collections rooms and/or storage cabinets are locked when reasonably possible and specimens on display are in secure cases, supervised by staff or closed circuit TV.
9	Written security procedures and guidelines are enforced and communicated to all staff, including maintenance staff and contractors.	Procedures are in place to control access to collections areas by contractors, visitors and staff. For example staff responsible for collections are notified of any works taking place in collection areas.
10	All collection and exhibition spaces are regularly cleaned by staff following sets of institutionally agreed procedures and methods that minimise risks to the collections.	Collections are kept free from dust accumulation and, for example, wood shavings created by maintenance work are immediately cleared away. Cleaning procedures avoid any chemicals that might harm the collections and avoid use of water in stores where humidity is controlled.
11	Environment reports (including integrated pest management) are used by those managing collections storage buildings to ensure that risks to the collections are minimised when any improvements or repairs are made.	Written information about the temperature, humidity and other requirements of the collections is available and used in decision-making processes when the area housing the collections is being modified.
12	There is a written emergency preparedness plan for the collections which includes procedures for dealing with disasters and the salvage of collections. All staff are aware of it and know their role within it.	In the event of an emergency such as a flood or a fire, the institution has a plan on how to minimise damage to the collections, for example how to treat waterlogged specimens to avoid further damage. Ideally local emergency services have been advised and are engaged in emergency planning and there is a review, training and testing mechanism.

(*continued*)

Table 3.3 Cont.

Collection Management		
This section aims to identify what procedures are in place to ensure proper management and documentation of collections		
	Question	*Explanation*
1	There are procedures in place to ensure that the collections management policy is being implemented.	This includes procedures for: accessioning/deaccessioning; acquiring; appraising; assessing significance; lending and borrowing.
2	Records are maintained on accessioning and deaccessioning of all specimens.	Information is compiled and kept up to date on new items being added to the collection or items being disposed of. For example details of donor/ collector, description of item(s), date of donation/disposal.
3	Each accession is marked or labelled with a unique identifier, which is produced to agreed standards and does not damage or deface the object. Previous markings/labels are preserved.	An accession can be an individual item (such as a giraffe) or a group of items (such as a tray of insects). Each accession should have a permanent code associated with it.
4	A copy of the accession register/ catalogue is stored securely at a separate location.	Copies of registers (paper or electronic) are stored off-site to ensure that the information is preserved in the event of a disaster.
5	There are procedures in place to ensure that any acquisition of new collections/specimens is in line with the institution's priorities and policies.	Staff are aware of the institution's priorities for accepting new collections. For example the institution has agreed which regions, taxa, quality of material it will accept or collect. Ideally these should be written.
6	There are written procedures for specimen loans and returns and they are consistently applied (when loaning specimens from the collection and when borrowing specimens from another institution).	Staff have access to documents which set out procedures for internal loans (such as loans for exhibition/ research within the same institution) and external loans. Ideally there is a computerised system in place for all departments.
7	There is a written procedure regarding the internal movement of specimens and all movements are recorded.	Internal movement is considered to be the movement of an item to a different room or building rather than a shelf in the same room. Staff have access to the rules regarding internal movement of specimens and movements should be documented. For example record on a database, or a tag placed in the position where the item originated.

Table 3.3 Cont.

Collection Management		
8	There is equipment and written guidelines available to staff and visitors for safe handling and transportation of specimens, specimen containers and conservation materials.	There is minimum risk to users and managers of collections. For example instructions are available and instruction given for use of trolleys, specialist lifting equipment, ladders and hazardous materials.
9	All collections data is stored in, and accessible via an electronic collection management system.	Data may be stored centrally or at departmental level. Ideally data is backed up and stored at a separate location and level of access is determined by institution, such as who can have access and when.
10	Advice is available on legal aspects of collections management.	Advice can be sourced either internally or externally. For example intellectual property rights, Access and Benefit Sharing (ABS), acquisition and disposal.
11	At least 75% of the collections are sufficiently ordered to permit staff to locate material easily.	This does not apply to all staff, but those responsible for managing the collection.

Collection care and preparation

This section aims to identify which collection care procedures, for example remedial and preventative conservation, interventive processes and storage are in place to ensure that collections care standards are adhered to.

	Question	*Explanation*
1	There are procedures in place to ensure that the collections care policy is being implemented.	Policy refers to what can and cannot be done at an institutional level. Procedures describe how the policy is implemented. In this case procedures include: condition checking; conserving; controlling pests; emergency planning; ensuring health and safety; handling/moving; insuring (including indemnity); managing collections care information; maintaining technical viability; monitoring and controlling environment; risk assessing and managing; packing.

(*continued*)

Table 3.3 Cont.

Collection care and preparation		
2	All conservation work is carried out according to recognised standards.	Conservation standards can be institutional, international or specific to types of collection. For example ECO, Icon ethical guidelines. This applies to anyone doing conservation work (such as staff, visitors, volunteers and contractors).
3	Collections are periodically surveyed to assess their condition and deterioration and to identify any risks. The results are used as part of the planning process.	
4	Conservation requirements have been assessed and prioritised in consultation with relevant staff and management.	Data from surveys (see Question 3) is used to prioritise conservation actions.
5	Any interventive processes undertaken on an item are documented. This includes an assessment of any loss or change to the item that will be caused.	Any interaction with an item which results to a change in it should be recorded. Interventive processes include conservation, preparation, replication and destructive sampling.
6	The collections environment (storage and display) is controlled within agreed environmental parameters. The environment is monitored and an early warning detection system is located in high-risk areas.	The environment in which collections are stored or displayed (such as temperature, relative humidity, light) is managed. Ideally environmental parameters should be written down and accessible. Examples of high risk include floods and alcohol leaks.
7	Technical support is available to facilitate collection care.	There is support for activities such as specimen preparation (for example plant mounting) and operating specialist conservation equipment.
8	Facilities are available to the institution to ensure the conservation of the collections.	Institutions have access to appropriate equipment, working space and lighting for conservation work.
9	Procedures are in place to prepare items entering the institution. Preparation facilities are available (either within or outside the department).	Staff know how to process incoming material (for example by storing in alcohol, drying or mounting) to ensure that the item is preserved and have the appropriate equipment to do so.
10	Materials used to prepare and store collections minimise deterioration to the collections and their associated information.	Materials should all be tested and be of conservation grade (using a suitable conservation test, for example the Oddy test, for safety of materials to be used in and around art objects) and suitable for purpose.

Table 3.3 Cont.

Collection care and preparation		
11	There is an established standard for storage furniture. There is a review and assessment of current storage furniture and future requirements.	The storage standard can be institutional, international or specific to types of collection. For example the material furniture is made of must not be detrimental to the collections, cabinet door seals must be pest proof.

Access to collections		
The aim of this section is to identify what the department does to make its collections accessible to users while minimising risk to items and people.		
	Question	*Explanation*
1	All collections are accessible to internal and external users via loan or visit.	A user can view, use and measure the collection (object and/or image) either by loan or visit.
2	Access to collection storage areas is restricted to authorised persons.	Authorised persons can vary depending on local conditions but would include restrictions such as only staff responsible for a particular collection can gain regular access to those storage areas. Also visitors may need permission from collections staff to gain access to the collections. For example collections staff might give visitors and other users authorisation in the form of an electronic entry pass, a signed-out key or entry code.
3	Response to a request for access occurs within a specified time.	All loans or enquires to visit the collections are processed within a timeframe set by the institution.
4	Information on visitors to collections is recorded.	The institution identifies people specifically using the collection. For example recording (at minimum) their name, institution and address.

(*continued*)

Table 3.3 Cont.

Access to collections		
5	There are written procedures for controlling access to the collections by visitors.	Staff are aware of procedures for managing visitors. For example recording personal details and restrictions on access.
6	Instructions on handling collections, and related security and health and safety issues are supplied to visitors.	Information is provided on how to handle the collections in such a way to minimise damage and loss to the collection and ensure personal safety. This should comply with institutional regulations and ideally be in writing.
7	Users of the collections are made aware of the laws and regulations relating to their use.	Any restrictions on use of collections are made known to users. For example those relating to Access and Benefit Sharing.
8	The institution has a written policy on the terms and conditions of use of collections.	The policy refers to what can and cannot be done at an institutional level. Procedures (as in questions above) describe how the policy is implemented. For example a policy on whether destructive sampling is allowed, or returning any data captured to the collections owner.
9	Collections are organised by a recognised system allowing items to be easily located by users.	A system is in place that allows users unfamiliar with the collection to locate a particular item easily.
10	Visitors are provided with a suitable work space that minimises risk of damage or loss to collection.	The areas provided for visitors do not result in damage to collections.
11	Staff are available to support access to the collections.	Staff help provide physical access to collections, prepare loans for visitors and offer technical support.
12	Associated collections data (for example field notes, card indexes, databases) are available and accessible to visitors.	Data such as field notes, old labels, scans, images or illustrations that are not obviously visible from the physical specimen are accessible to visitors.
13	General information on the collections is available on the internet.	Users should be able to find out, in general terms, what the most significant collections the institution holds are. For example "Birds", "Permian fossils", "South American Solanaceae", "Cornish minerals".

Table 3.3 Cont.

Access to collections		
14	Digitisation priorities for collections and associated collections data (such as field notes) have been identified.	There is a planned (rather than unplanned) programme for taking and making available images of the collections.
15	There is adequate space and equipment for visitors to achieve the aims of their visit.	There are sufficient desks, bench areas and facilities, such as with microscopes, to allow visitors to work with collections.

Molecular Collections

The aim of this section is to assess how an institution is managing those collections specifically intended for molecular research, such as DNA, tissues and proteins. These questions are aimed at institutes with biodiversity collections such as museums and herbaria who store molecular collections and particularly where they are under the responsibility of non-specialist collections staff as part of their remit. It is not aimed at specialist molecular facilities such as biobanks and medical collections, although there are overlaps. Increasingly, institutions such as museums and herbaria are building specialist facilities managed by specialist staff and these will align more closely with the biobank community. In these cases refer to organisations such as ISBER (www.isber.org) who have self-assessments specific to that level of facility.

	Question	*Explanation*
1	The institution has written policies and procedures governing the management and use of molecular collections and all users are aware of them.	This covers areas such as acquisition, storage, accessioning, databasing (terms and conditions of use), linkage to voucher material, health and safety.
2	Storage conditions are at a standard to minimise degradation of molecular components.	Different storage facilities are needed for different types of material (for example leaf material and extracted DNA). Material is stored under the appropriate cost-efficient conditions.
3	The institution has access to and consults specialist advice on molecular collections when needed.	This might be expert networks such as ISBER and BBGB, technical expertise, service and maintenance, relevant publications.
4	Facilities are available which allow the use of the collections without posing a health risk to users.	For example, gloves suitable to be used at -80 degrees, down flow benches for silica gel use, adequately ventilated rooms for liquid nitrogen use.

(*continued*)

Table 3.3 Cont.

Molecular Collections		
5	An efficient local inventory storage system is in place which allows material to be easily located.	Linking physical samples to databases for rapid sample retrieval and space-saving sample deposition, for example FreezerPro/Works.
6	All sequences and other data obtained from the collections are linked by unique identifiers to both samples and to any museum/herbarium specimens from which samples were taken. All samples taken and sent to other institutions are trackable.	For example, there is always a means to track where a DNA sample taken from a specimen has been sent and under what conditions of use. This is usually done through unique identifiers and management systems (see Question 7).
7	Accessible Laboratory Information Management System (LIMS) linked to collections management metadata.	For example, collectors' field notes and a herbarium specimen are clearly linked to a DNA accession. This may be by using a unique identifier and database, and compatible shared data portals for enhanced discoverability and accessibility.
8	There is a disaster plan in place to enable samples and specimens to be protected from loss or damage through thawing out due to emergency situations such as power outages and equipment failure. All relevant staff are trained and have practiced these procedures.	For example, there are empty back-up freezers available to receive material from failed freezers; freezers can easily be removed from an area under threat of flooding or to a new power source, emergency power supplies are available. Staff do regular test runs of the disaster plan.
9	All terms and conditions agreed at the collection and acquisition stage regarding use of material are accessible and observed. Use of material and where the resultant data/material is subsequently deposited is recorded and tracked.	All material should have permit information associated with it and any users are expected to sign an agreement to abide by conditions such as might be required by the Convention on Biodiversity Nagoya Protocol on Access and Benefit Sharing. These may comprise suites of Prior Informed Consent (PIC), Mutually Agreed Terms (MAT) and Material Transfer Agreements (MTAs).
10	Depletion of destructively sampled molecular collections is tracked to ensure they are not exhausted.	For example, destructive sampling of silica dried leaf material requires permission and amounts remaining monitored. System triggers an alert at minimal levels remaining. Extracted DNA is multiplied as stocks become low.

Table 3.3 Cont.

Molecular Collections		
11	There are guidelines in place to help future-proof use of molecular collections.	These are to make sure that any treatments or sampling carried out do not affect any possible future uses as far as is predictable For example, there are guidelines on how much material to collect and information on the best in-field storage methods. Decisions are made on minimum numbers of duplicates to represent genetic diversity, cater for expected future "market" use and/or minimise repository storage costs (3 duplicates minimum: in use, master stock and back up sample).
12	An assessment of the quality of the material being accessioned for long term storage is routinely made and recorded.	For example, is a DNA extraction of sufficient quality to be of use and are there quality control and quality assessment procedures in place and regular audits carried out? Records are kept of sample history such as environmental storage temperature records including during disasters such as power outages.

Digital Collections		
The aim of this section is to assess how an institution is managing its digital collections and associated data.		
	Question	*Explanation*
1	The institution has written policies and procedures governing the management of its digital collections and users are aware of them.	This would cover areas such as acquisition, storage, accessioning, databasing (terms and conditions of use), linkage to voucher material, health and safety.
2	All data and metadata are recorded to accepted standards such as Dublin Core, Darwin Core.	
3	Digital collections are easily accessible unless particular restrictions are in place.	This may be as an image on the web or a high-quality download. Sensitive data could be information such as locality of a rare species, which may attract collectors and result in possible extinction or risk of theft of high profile specimens such as rhino horn.

(*continued*)

Table 3.3 Cont.

Digital Collections		
4	Suitable workstations and facilities are available for the digitisation of material which does not damage the collections or pose a health risk to users.	Work spaces have enough room for specimens to be manipulated and put down without risk of damage. Scanners and cameras do not touch specimens in ways that might cause damage. Supports ensure that specimens are supported adequately and not subject to undue stress.
5	Digital collections and associated metadata (such as details of photographer, crowdsourcing information, date) are regularly archived.	These backups should be made in a widely accessible format and periodically checked for technological developments. Ideally RAW files should be stored as well as a range of other formats such as JPG, TIFF.
6	There are regular discussions to make sure the digital collections continue to be in a format suitable to future access in light of new software developments.	Regular meetings take place to keep up to date with software developments and to develop strategies to prevent obsolescence.
7	There is a quality control mechanism in place for digital collections and associated data.	For example, is the image in focus, does it have a scale bar, has it been cropped correctly, is all information visible and correct (for example readings from OCR technology)?
8	Terms and conditions of use are clearly displayed and observed.	For example, intellectual property rights, copyright, Access and Benefit Sharing (ABS) requirements.
9	Digital collections are submitted to an international aggregator such as GBIF (www.gbif.org).	This will increase access and improve awareness of collections.

Source: www.synthesys.info/network-activities/synthesys3-na2/self-assessment.html

APPENDIX 3.2: EXAMPLE OF A HIERARCHICAL SCHEMA FOR A PREVENTIVE CONSERVATION STRATEGY

This schema is a section of the Smithsonian National Natural History Museum's Collection Stewardship Plan. It leverages the information gathered and maintained in the Collections Assessment and in Collections Space Survey.

Purpose of the schema

To guide the prioritisation of activities, spending and funding requests consistent with the museum's strategic plan and stewardship responsibilities, while maintaining flexibility should unforeseen opportunities or needs arise.

Expected outcomes

- Improved collections care.
- Clear guidance that ensures that actions and spending fit with a strategic approach to physical care of the collections.
- More easily tracked outcomes for reporting purposes.
- More easily developed funding requests for internal and external proposals.
- Aligning funding and labour provided with the organization's capacity to use those resources effectively and efficiently.
- Better understanding among staff for the basis for prioritisation.

Guiding principles for actions

- Minimise the potential for loss by using a "prevention" mind-set.
- Bring all collections to acceptable levels before "polishing" a few.
- Fix a problem once.
- Share and collaborate where possible.
- Demonstrate that resources (space, funds, time) are used efficiently and effectively.

Schema

This schema is built upon the assumption that prevention is fundamental to long term preservation. In general, prevention takes increasingly specific forms that are ranked here as levels:

Level 1 Appropriate building, with climate, light and access control.
Level 2 Appropriate space divisions within the building.
Level 3 Container/cabinet is appropriate and in good condition.
Level 4 Housing supplies such as boxes, papers, foams and vials or bottles are appropriate.
Level 5 Addressing the physical condition of the item itself, including ensuring that it is not overcrowded.

The schema uses collections assessment data about condition and importance to guide prioritisation of projects.

A. LEVEL 1: BUILDING or LARGE SPACE
   Collections are housed in multiple spaces in two large facilities. For the purposes of the Physical Care Plan, collections spaces include exhibition and educational spaces where collections are placed. While the majority of collections are on-site, a discrete number of collections are maintained elsewhere via Off-Site Enhancement agreements and tens of thousands of specimens are on loan.

Level 1 actions should address:

- building envelope;
- security;
- housekeeping;
- environment/integrated pest management;
- safety;
- training;
- risks of inaction;
- traffic through the space;
- adjacencies.

Example Level 1 priority actions

- Design and construct an exhibition hall that can consistently meet current environmental standards.
- Hire a museum conservator to assist with collections problem solving and risk management at the museum level, including liaison with building management groups.
- Renovate spaces and move target collections into them, in new cabinetry using appropriate supplies.
- Establish a museum integrated pest management committee.

B. LEVEL 2: SPACE

As spaces are renovated, collections are placed behind walls whenever feasible. This ensures that the two-barrier security standard is met while providing some protection against pests and environmental fluctuations. Other collections are in unrenovated spaces and thus are not separated from offices, foot traffic and other functions. This means that some collections are in hallways or open ranges.

Level 2 actions should address:

- room or hall;
- security;
- safety;
- housekeeping;
- environment;
- arrangement (ease of access);
- arrangement (efficient and safe use of space);
- training;
- risks of inaction;
- evaluation of compactor-capable spaces.

Example Level 2 priority actions

- Labelling of collections –based on hazards in each space (i.e., room).
- Clearing attics of collections items.
- Plan for collections-related documents currently in the attics.
- Relocation of fish collection – in progress.

- Labelling cabinets for contents.
- Consistent IPM risk zoning.
- Wayfinding signs (with owner/manager/steward of space).
- Walkthroughs with Fire Department.
- Collections space database update.
- Permanent common swing space (with standardised cabinets).

C. LEVEL 3: CABINET or SHELF
Some facilities are, or soon will be, undergoing various renovations. The museum has committed to replacing outdated storage cabinetry and shelving when the spaces themselves are renovated. This view of an appropriate collections "house" is a major step forward to ensuring that the collections' physical conditions are optimal. Because the funds for storage cabinetry and associated supplies are provided by external sources when collections are affected or moved as result of building renovation or construction, the museum has been making the replacement of relevant equipment and supplies a very low priority for internal funds *unless* the collections items themselves are in imminent danger of damage or loss.

Level 3 actions should address or consider:

- assessment data about storage conditions;
- quality and number of cabinets;
- re-use or disposal of cabinets, costs;
- verifying accurate reporting on use of space
- non-museum sources of funds;
- inspection of new equipment;
- arrangement (ease of access; efficient use of cabinet/shelf space);
- training;
- dependencies affecting intended actions;
- beginning to explore the standards and spaces for archival and collections-related documents;
- documented specifications for standardised cabinets;
- including exhibition and education cases as cabinets;
- connect to case mapping plans.

Example Level 3 priority actions (spread over multiple years)

- Comprehensive list of cabinets per department with count to be replaced.
- Map of cabinet placement in Building 1.
- Standardising delivery and installation of new cabinets.
- Cabinet replacement in two collections.
- Special environment vault remediation – tentative solution identified move to level one.
- Microclimates for two specific collections.

D. LEVEL 4: INTRA CABINET/HOUSING MATERIAL
Collections that have been moved recently are most likely to be housed in appropriate materials, or what was considered to be appropriate at

the time. Some of these materials need to be reviewed or tested. Other collections need to be rehoused or undergo some kind of remediation. Still others need to be rehoused, or could benefit from a different arrangement or housing approach. A routine materials testing approach would benefit the collections, as would a more consistent inspection protocol.

Level 4 actions should address or consider:

- assessment data for STORAGE, ACCESS and IMPORTANCE criteria;
- quality and number of cabinets;
- re-use or disposal of cabinets, costs;
- types of materials in use (containers, supports, trays);
- container design with regard to standardisation within drawers (unit tray system);
- non-museum sources of funds;
- inspection of new equipment;
- arrangement (ease of access; efficient use of cabinet/shelf space);
- training;
- dependencies affecting intended actions;
- testing of materials; arrangement;
- opportunities for leveraging purchases;
- workflow characterisation; opportunities for leveraging activities;
- risks of inaction;
- IPM treatment and inspection;
- interaction with plans for collection digitisation;
- centralised procurement for collections-adjacent materials and supplies;
- ability of computer systems to support specimen or location labelling.

Example Level 4 priority actions (spread over multiple years)

- Mercury capture project – in progress; long term project.
- Air exchange monitoring project.
- Naphthalene remediation.

E. LEVEL 5: ITEM

The Smithsonian National Museum of Natural History usually focusses on caring for collections in the aggregate rather than at the item (specimen or object) level. Occasionally an item warrants individual treatment in light of its importance, near term intended use or the feasibility of achieving a benefit for the costs expended. Treatment of items prior to exhibition would fall into the Level 5 category, as would treatment of specific ethnographic objects in collaboration with traditional communities.

Level 5 actions should address:

- assessment data for IMPORTANCE criteria;
- digitisation as a form of documentation and preservation;

- exhibition use;
- reparable damage;
- feasibility (cost/benefit);
- resources needed (funds, time, space, equipment);
- risks of inaction;
- training;
- security.

Example Level 5 priority actions

- Repair of earthquake-damaged ceramics.
- Inspection and care of items removed from exhibit.
- Standardised format for tags to indicate temporary removal (also known as "Drop Slips").

APPENDIX 3.3: AN EXAMPLE OF AN ACTION PLAN FOR A LEGAL DOCUMENTATION PROJECT

Project purpose: to ensure that the collections held by the museum were and will be acquired legally and to maintain records that document the acquisition process and decision making.
Vision of future state: all records are fully processed. There is no pending paperwork.

The project team met to discuss the current situation, their skills and resources, and the desired future state of the collection's legal documentation. They conducted a brief "SWOT" analysis, to understand Strengths, Weaknesses, Opportunities and Threats associated with the situation.

SWOT

Strengths:

Knowledgeable team.
Collections information management system is robust.
Cooperative curators.
Support from museum leadership.

Weaknesses:

Variability in what data are used in record keeping system.
Slow processing of documents by staff.
No way of holding staff accountable.
No unified system or standards for workflow or data.

Opportunities:

Chance to develop good process.
Funding available for student labour.

Threats:

Loss of data.
Illegality, legal liability.
Reputational risk.
Inability to use collections because of undocumented terms of use at acquisition.

Strategic plan

After considering the state of the collection's documentation and their SWOT findings, the team agreed on the following strategic plan. They would:

- understand variability of different types of paperwork for different types of collection items;
- develop workflows that will process work promptly;
- assess data and files systems that will hold information properly;
- train staff;
- establish schedules;
- monitor work;
- review progress;
- announce and celebrate improvements;
- reassess status, and continue the strategic cycle of work.

The team set the following tactics or action steps in priority order, to be completed within 18 months. The Collections Leader is accountable, the Registration team is responsible, the Curator is consulted, and the Department is informed.

- identify necessary elements for documentation;
- agree on what are acceptable forms that information can take: signed letter, email, digital signature on digital doc, etc;
- gather scenarios on how the collections can come into the museum and when they are likely to have data or be missing data;
- develop workflows;
- test workflows;
- train staff on workflows;
- set schedules for processing;
- manage work per schedules;
- review unprocessed acquisition papers;
- identify missing elements;
- determine whether missing parts can be acquired or whether alternate sources of information will be required;
- report on progress; celebrate achievement;
- review processes, make adjustments as necessary and establish new workflow to avoid non-standard work processes.

4 Governance, legal aspects and policy relating to strategic management of natural science collections

Introduction

In the previous chapter we looked at the various elements of the planning process, including gathering data to support your goal, and ensuring you have the people you need in your team to plan and achieve your goal before developing an action plan to make it happen. However all this could be in vain if what you are planning to change does not comply with the laws of your country, international regulations or the balances, controls and policies within your institution. In this chapter we look at these in more detail in the context of governance.

Governance, whether it takes the form of laws and regulations, decisions by boards of trustees or review committees, or policy is necessary for organising and managing the work of the museum. Simply put, governance is the array of laws, regulations, policies, practices, rules and procedures that tell you what you *must* do, what you *may* do and what you *may not* do. Clear and informed governance is a key element in leading a museum to success with its purpose and plans and to managing risks to collections, staff and institutional sustainability and reputation. Museums ought to know the legal basis for their existence: if they were established by law or decree, the museum must comply with that law or decree. If the museum was established via other means, those means will contribute to the governance of the museum. In any regard, museums should know what entities control their existence, resources and activities. Museum staff benefit from understanding levels of authority and responsibility, how to get clarification and guidance on governance matters and how to apply governance documents such as regulations and policies. This chapter addresses questions about use of policy and the role of governance in strategic decision making, but not about how to write policy as there are excellent resources on that topic (see Simmons, 2018).

Questions

- How is governance used in collections organisations?
- What are the risks associated with governance and how can you manage them?
- How can I use the power of policy and governance to minimise risks and help me achieve my goals?
- Is it possible to change a governance framework?

How is governance used in collections organisations?

Governance sets the framework for decision making and autonomy. All institutions sit within some kind of legal framework, whether in relation to the nation, the state, a university or research centre, as led by a board of trustees, for example. That framework could be expressed as a royal decree establishing the museum, a law, the charter establishing a university, the terms of a bequest that provides the funding that starts a museum and others.

Exercise

Thinking about an institution that you are familiar with, ask yourself:

- What is its legal framework?
- How are significant changes or decisions approved? Who reviews plans and proposals that require resources?

All staff have the obligation to understand what the governance is for their organisation. Each person should know what actions they make take, based upon their defined role within the organisation. They need to know, for example, if they may purchase an expensive mineral specimen from the open market without prior approval, or if the purchase must be authorised. Generally, such roles are expressed in job descriptions or recruitment documents and they are embedded in the structure of an organisation (see Chapter 5). Staff benefit from knowing how resources are obtained, allocated and controlled, how the larger fate of the organisation is determined, how staff are hired or removed and other aspects of management. Knowledge of these aspects reduces questions and uncertainty when considering routine or controversial actions, making staff more confident and capable.

The hierarchical nature of governance

Laws and decrees set down the rules for what is permissible within the nation or state. Regulations implement laws and decrees. Policy implements the vision, values and professional standards of the museum and expresses how a museum complies with regulations. In this way, policy serves as a starting point for many strategic decisions. The hierarchy is as follows:

Laws/decrees ➔ regulations ➔ POLICY ➔ procedure

Which is the most rigorous: laws or ethics?

Law indicates what you can and cannot do legally. Ethics are a series of agreed behaviours that express shared values within a legal framework. If followed, ethical practices will lead to the best possible outcome for most people. If you imagine a controversial situation, you may know what actions are legal but be unsure about actions that are ethical. What is the essential thing to do, comply with the law or meet the ethical standard? Ethical standards are usually more rigorous than laws: many have said that the legal standard is the lowest standard and that museums should be acting at an ethical standard. The simple answer is to strive to always meet the ethical standard as it incorporates acting legally.

Example: Legal, ethical or both?

Law

Museum M will only use legal methods to acquire collections (compliance with the law). In addition, it will not knowingly borrow specimens that may have been illegally acquired (meeting ethical standards).

Ethics

Many professionals have domestic partners who are also experts in the same line of work. Sometimes questions of nepotism arise when a partner or family member is among the applicants who are qualified to fill a job vacancy. It may be legally acceptable to hire your spouse to do work, but it usually is not considered ethical because doing so could imply that you have not followed standard hiring practices, among them considering all other candidates fairly and offering a fair standard wage that is neither inflated or unduly low.

Most professional societies have ethics statements and codes of conduct, and increasingly they set forth codes of conduct for professional meetings. Among these, see the American Alliance of Museum's website on ethics, standards and professional practices at www.aam-us.org/programs/ethics-standards-and-professional-practices/core-standards-for-museums/. ICOM provides a code of ethics specific to natural history museums and this can be found at https://icom.museum/wp-content/uploads/2018/07/nathcode_ethics_en.pdf.

The Global Biodiversity Information Facilities' code of conduct for conferences can be found at www.gbif.org/code-of-conduct.

What are the risks associated with governance and how can you manage them?

Everyone who manages resources or makes significant decisions needs to know the laws that cover their collections at the local, state, national and international levels. Staying current on applicable laws is a routine task that should be planned into the work of the museum, or, for example a service provided by resident legal advisors either on a retainer or available on a fee-for-service basis. You may also gain information through professional societies. Knowing the rules that apply to your work at your level of authority and responsibility, and knowing what you may and may not do, is a significant way of mitigating risks of infringement and damaging penalties. Ignorance of governance matters is not an acceptable reason for infringement. Penalties can be high – financial, reputational, ability to do future work, increased scrutiny or other unwanted results.

Lack of staff awareness of the law can be a major institutional risk leading to a chain of consequences, impacts and a need for mitigation as follows. For example:

- Risk – staff are unaware of the laws that cover their collections.
- Consequences – collections are disposed of contrary to laws on disposal of museum objects or staff purchased illegally obtained items.
- Impacts – legal action, damage to institutional reputation, restrictions on future actions.
- Mitigation – communication with enforcement agencies or injured parties, staff training, revised procedures.

Example: Planning for permits: a hard lesson

A museum in country A was collaborating with artists from country B to develop an exhibit on country B's traditional art. Typical materials used in the traditional art included rocks, minerals, ivory, bone, animal skin and hair and other natural and man-made products. A curator from the museum had not planned ahead to obtain a permit for the importation of ivory and animal skin and thus could not legally receive works of art containing those materials. The opening date for the exhibit was rapidly approaching and the artist quickly packed and shipped the art works, including items that should have been, but were not, covered by a permit. When the customs inspector in country A inspected the shipment, he identified the art works that could not be received legally without a permit. The customs officer followed the proper procedures and contacted the relevant law enforcement agency. The head of the enforcement agency called the director, not the head of collections, at the museum. The director learned that the museum had violated laws and was at risk of losing existing permits that allow the shipment of regulated animals and their parts. The director and the enforcement agency discussed the potential outcomes of the situation. Ultimately, the museum refused the shipment of the unpermitted items and they were returned to the sender in country B. The exhibit was missing some items that helped to tell the intended story, but the museum preserved its ability to legally import future collections.

Sample areas of legal concern

Legal aspects touch on many areas of our work in collections. Here are some examples that pose risks of legal infringements and that should be kept in mind when planning your work or projects.

Ownership or rights regarding use (access and benefit sharing, copyright, open access data)

When many of today's museum professionals began their work, the concept of "ownership" was very clear: items donated or acquired by the museum would belong to the museum with "no strings attached", meaning that the museum could use the items and their associated data in any way that was desired, for the indefinite future. Sometimes this was expressed as "full and clear title", with all rights to the item being conveyed and no limitations or conditions regarding ownership and usage that accompanied the items. Museums preferred this kind of ownership for good reason. They avoided conditions that restrict the use of items to certain people or groups, limitations on whether items could be transferred or sold, destructively sampled, used for commercial purposes and so on. Museums typically avoided restrictions that required that an item be on permanent exhibit for reasons that can be understood given the long life of most museums. Today, the ownership concept is far more nuanced because museums often cannot acquire items without agreeing to some kind of restrictions. With increasing frequency, new acquisitions are conveyed without the full rights we received in the past. Now there may be limitations on analytical uses, limitations on who may view or use an item or requirements that any benefits derived from study of the item be shared with the source country. For biological collections items, many researchers and museum staff attribute or blame these kinds of limitations on the Convention of Biodiversity's Nagoya Protocol, but that is not fully justified. The Nagoya Protocol has not granted new rights to countries over the biodiversity within their borders; countries are exerting rights they already have (see Chapter 6).

How can museum staff prepare for potential controversies about ownership and rights? Internal discussions are helpful when they occur among researchers, collections managers and directors or other leadership that explore what kinds of rights the museum feels it must have, those that can be decided on a case by case basis and those that present low impact or little risk. Museums can also discuss their relationshops with traditional or source communities, working towards a new definition of ownership. It is always easier to begin such discussions when the subject is hypothetical, before anyone has a vested interest in a specific outcome. Once the general decisions are made, they should be written as policy and widely distributed to staff. The information should also be reflected in registration systems and documents that convey ownership and rights.

Shipping/collections transit regulations

Currently, shipping formalin or alcohol-preserved collections via air or ground requires that specific procedures be followed. These apply to the amount of fluid that can be present and the containers that can be used. These regulations can change, so it is wise to stay current with the national postal codes, and consult the International Air Transport Association regulations (www.iata.org). Compliance with shipping regulations reduces the risk of damage or loss, and helps to ensure that items make it safely to the intended destination, on time and without additional costs.

Prior Informed Consent (PIC) and Mutually Agreed Terms (MAT)

Many countries are now requiring that formal documents be signed that contain agreements on how acquired collections may be used.

Hazards inherent in specimens or resulting from materials applied to specimens

It is well known that some natural science collections have inherent qualities that pose risk, such as naturally occurring radioactivity, or have acquired qualities that pose risk. Acquired risks may be the result of treatment, such as the use of arsenic or mercuric salts as a preservative, and other toxins used to reduce or eliminate pest infestation. These risks are well covered in *Health and Safety for Museum Professionals* (Hawks et al., 2011), among other texts. The point here is that everyone should understand the legal implications of hazards in the collections. The legal standard in some countries may indicate that users of collections must be provided with information about hazards and personal protective equipment such as gloves or goggles. The ethical standard could include the added emphasis on proactive notification about risks to collections users who are in their reproductive years so that they can make informed choices about exposure to hazards.

Table 4.1 Example of the hierarchy of governance of bird specimens in the United States

Topic	*Law*	*Regulation*	*Policy*	*Procedure*
Acquisition (in the US, of bird specimens)	The Migratory Bird Treaty Act.	Migratory Bird Hunting Regulations which are published annually in the Federal Register.	Museum M only acquires specimens that were collected legally.	Museum M's collectors must provide documentation of collecting activity that includes any relevant permits and collecting dates. All such documentation will explicitly be linked to specimens and will be maintained in permanent record systems.

How is policy distinct from law, regulation and procedures?

Using a chart or diagram to understand the relationship of the many elements that guide museum work can be very helpful, especially when working with staff new to the museum profession. The distinction between law and regulation is usually clear as it can be verified through publicly accessible resources. The distinction between policy and procedure may seem more confusing, and as it is established at the organisational level it may be more difficult to clarify. Simply put, policy states what you will or will not do, and procedure explains the steps that ensure that policy has been met. We have provided an example in Table 4.1.

Exercise

How would you fill in the chart for "loans" in Table 4.1?

How can I use the power of policy and governance to minimise risks and help me achieve my goals?

Policy and other aspects of governance guide the action of a museum. When policies are clear, well informed and relevant they serve a dual purpose of protecting the museum from unwanted or undesirable actions while supporting what is needed and desirable. Many staff have the experience at some point in their career of being asked – or told – by the management of the museum to take an action thought to be inappropriate or distasteful. How can staff disagree or advocate for a different action?

Exercise

Imagine you have been asked to dispose of a collection of minerals and you disagree with your director's intentions. Consider:

- Is the action legal?
- Is the action permissible or acknowledged as part of the museum's charter or enabling legislation, or written statement of purpose?
- Is the action considered unethical per the terms of any relevant professional societies?
- Are there community standards that apply to the proposed action?
- Can you identify clear benefits or liabilities that will result from the action?
- Is the action within your own level of authority and responsibility?
- Can the action be correctly completed with available resources?
- Will taking action necessarily remove the possibility of future opportunities?

Where policy can help protect you and your collections from unwanted action

Illegal acquisition

Museum M's collections management policy clearly states that all acquired specimens must have been collected legally. Biological, anthropological or geological specimens that were illegally collected will not be accepted. When Donor D offered a tiger skin to Museum M's director, the director asked the registrar to meet with Donor D. The registrar learned that Donor D got the tiger skin from a friend who went on a group hunting trip to Asia. The group of hunters had been authorised to hunt common animals and to keep their skins. The hunter's friend wanted a tiger and he killed one without authorisation. Donor D told this story to the registrar, but he could not give the registrar any documentation that would show the tiger was killed, exported and imported legally. The registrar recounted what had been learned to the director who then decided to decline the donation.

Out of scope acquisition

Museum M's associate director asked the head of Department 2 to accept a large sculpture from a country in Asia. Museum M's collections acquisition policy requires that each department have a collecting plan that guides what will and will not be added to the collections. Department 2 did not want the sculpture. They did not have any collections from Asia and in fact had already developed a collections plan that explicitly focussed all new acquisitions on specimens from Africa. Department 2 declined to accept the large Asian sculpture. They relied upon their policy and further explained other factors that contributed to their decisions: they had no acceptable place to house the sculpture, no research plans that would require study of the sculpture and no exhibition plans in the foreseeable future that would use the sculpture.

Use of hazardous chemicals

Museum P is dedicated to environmental responsibility and "green" work practices. Their stated policy is to use only passive environmental controls. They will not use toxins. When Researcher 7 found pests in a collection of animal skins, she insisted on the application of a legally regulated poison. Researcher 7 and the collection manager had a significant disagreement on what action to take. Researcher 7 said she would buy the poison and apply it herself, even though she did not have the required applicator's license. The leader of the department relied upon the environmental policy and state law about the poison to refuse Researcher 7's demands. Instead, the leader supported the collections manager's plan to follow the policy and place the affected specimens in a CO_2 bubble so that the pests and their eggs would be killed.

Untimely processing that hides problems

Researcher 22 decided to collect animals to supplement the collections at Museum M. She went into the field and although she had a valid export/import permit, she did not get a valid collecting permit for species X. Her collecting permit only covered specimens from species A, B, C and D. When Researcher 22 returned to Museum M and gave the shipment to the collection manager, she told the collection manager to place the shipment in a hidden corner of the space designated for pending collections and not to worry about registering and processing it. The collection manager knew that Museum M's policy required that new acquisitions be registered within 15 days, so he proceeded to register the collection. When he compared the new specimens with the collecting permit, he quickly found that species X was not covered. This meant that Museum M would not be able to acquire species X legally unless they could get an *ex post facto* permission from the country of origin. Without legal acquisition, species X would also not be eligible for research use by Museum M or by borrowers. After several meetings with the leadership of the museum, species X was returned to the authorities in the country of origin and Researcher 22 reprimanded for her actions.

Where policy can support or encourage actions that you want

Acknowledgement of hazardous materials leads to improved warning signage/notifications

Museum P acts upon its policy about environmental responsibility by assigning a staff member to survey the existing collections for hazards. Technician 3 and Conservator 1 find that many of the older vertebrate specimens have residual amounts of arsenic that cannot be removed. They investigate the standards for signage about this kind of poison and they place notification labels in the entry to the collections rooms alongside boxes of gloves so that collections users may take precautions.

Policy statements about accountability lead to regular shelf-checks and other inventories

Museum M's policy requires that at least two different collections are inventoried each year. The recent inventory of a collection that was last used five years previously initially showed that a valuable collection of baskets was missing – it wasn't in the cabinet indicated on the inventory. The next day, however, investigative work showed that the baskets were present, they had just been returned to the wrong cabinet by a staff member. It made the staff realise that the cabinets needed new labels, that they needed to be more careful and that regular shelf-checks are useful.

Acknowledgement about source country rights leads to improved information management

Museum M's policy states that they gain Prior Informed Consent (PIC) and they negotiate written Mutually Agreed Terms (MAT) before collecting in other countries. Museum M respects the terms they have agreed, so they must ensure that they can fulfil those terms. As per many recent collecting agreements, Museum M must ensure that those documents accompany the specimens and be honoured by subsequent users of the specimens. As a result, Museum M develops a more robust collections management system and new procedures that allow staff to scan, upload and link documents quickly and easily. This enhances the specimen records, as there is more information in the documents than is typically entered into a specimen record.

Acknowledgement about staff safety leads to compulsory field safety courses and budgeting for emergency response

Museum T values the safety and health of the staff and expresses these values in their policies, requiring the development of safety plans for equipment use and certain types of activities. After a field trip resulted in significant illness to a staff member who narrowly avoided death, Museum T's administration hired a consultant to develop and teach a field health and safety course. The course is compulsory for every expedition participant and must be renewed depending upon the environmental risks where the expedition will occur. Annual funds are set aside for emergency evacuation, if necessary.

How can policy help on a day-to-day basis?

Policy can help staff in their day-to-day work. An example is providing support for decisions on granting access when resources are limited or the person requesting access isn't a frequent user of the collection. Imagine that you decide to deny access. If policies are clearly written and availble for potential visitors and borrowers, preferably on the institution's website, then the requestor can more easily understand that the denial was based on policy. The person requesting to use the collection can then be certain that he or she is being treated equally to others, which can make refusal conversations easier and benefit institutional reputation. Policy has strategic impacts beyond the individual transaction.

Who can do what?

Many museums establish someone's authority at the point of hiring or when a significant change in duties occurs. Authorities and responsibilities can be indicated in job descriptions, or in policy or procedure documents that associate specific roles, e.g. director or curator with specific authorities (see Table 4.2). Some museums express authorities at different levels, such that the same phenomenon is handled with different levels of authority as

Table 4.2 Generic anonymous levels of authority for acquisition

Role	*Action*
Trustee	Approves purchases of new acquisitions that require large amounts of exhibit or storage space, or that require significant funds
Director	Approves purchases of new acquisitions that fit within the existing budget of the museum
Curator	Approves acceptance of donated items within their area of expertise
Staff	No approval authority

associated with different levels of risk, cost or resource use. If you understand your level of authority, you can propose strategic plans for yourself and your staff according to your level of control. Successful accomplishment of strategic plans may lead to new opportunities, new initiatives and professional growth for individuals and groups.

Is it possible to change a governance framework?

As can be seen from the hierarchical nature of governance, specific procedures are the easiest form of guidance to change because they indicate *how* a task is done. In many instances, procedures can, and should, change frequently so that they respond to changes in technology, staffing, resources and other factors. Policy, which is written in more general terms than the specifics of procedures, is enduring but should be reviewed on a regular basis so that it continues to be reflective of the vision and purpose of the organisation and is a response to changing legal requirements. Collections professionals have a distinct responsibility for establishing specific procedures regarding collections. They also bring important perspectives and expertise when participating in the development of museum policies. By participating in reviews, discussions and corrective or preventive action and revisions of policy, a collections professional can help shape levels of governance within the museum. Collections professionals who wish to contribute to development of laws and regulations, or their implementation, should consult the administration of their museum. In the US, some museums respond to requests for comments by the federal government for public opinion and recommendations on proposed regulations. Others choose to provide input on law and regulation via official comments made by professional societies. Both are appropriate ways of being engaged in the larger governance framework that affects the museum. Both can be effective in presenting the museum's perspectives on regulations to the law makers who may not know or understand the consequences of the regulation on the museum.

Progress in implementing or improving governance can be measured using performance indicators. For example:

- effective date and renewal dates for essential policies;
- milestones achieved in implementing policies;
- effective date and milestones achieved in strategic plans;
- number of staff trained and training dates for compliance with regulations and policies;
- number and penalties for infractions.

Summary

Recurring elements in this chapter include the importance of knowing the laws, regulations and policies that apply to the museum and one's level of authority and the importance of written documentation. While not explicitly stated, an underlying element is the need for informed, professional and constructive action by staff at all levels when confronted with controversy and disagreement. Courageous action is especially important when the controversy or disagreement originates with or includes persons having greater authority. In such instances, the power of written, broadly distributed and well-understood policy becomes even more important for the protection and progress of the organisation. It can be very challenging to disagree with colleagues and the administration or point out errors they have made. Relying on clear policy and procedure can ease those conversations because they allow everyone to focus on the policy and behaviour instead of the personalities and worth of persons involved. The more frequently and consistently a courageous and professional approach is used, the easier and the less necessary it becomes. Collections professionals should develop the habits of regular conversation about policy matters and questions so that questions and controversies are addressed promptly. Early input that corrects or avoids potential missteps, or that promotes desired outcomes, is always beneficial. Early and respectful input tends to be less embarrassing and damaging than deferred intervention – and less embarrassment when making a correction usually leads to better working relationships. Being known for constructive and professional interaction with colleagues when responding to governance matters builds confidence in the collections professional's abilities and enhances his or her credibility – and the entire museum will benefit.

References

Hawks, C.M., McCann, K.A., Makos, L., Goldberg, D., Hinkamp, D.C., Ertel, D., Jr., and Silence, P., (eds.), 2011, *Health and Safety for Museum Professionals*. New York: Society for the Preservation of Natural History Collections.

Simmons, J.E., 2018, *Things Great and Small: Collection Management Policies* (2nd edition). Lanham, Maryland: Rowman & Littlefield.

Further reading

Buck, R.A., and Gilmore, J.A. (eds.), 2010, *The New Museum Registration Methods* (5th edition). Washington: American Association of Museums.

Malaro, M.C., and DeAngelis, I.P., 2012, *A Legal Primer on Managing Museum Collections* (3rd edition). Washington: Smithsonian Books.

Yerkovich, S., 2016, *A Practical Guide to Museum Ethics*. Lanham, MD: Rowman & Littlefield.

5 Staff resources

Introduction

Natural science collections require expert input to ensure their long-term survival, maximise their accessibility and to add and update the information associated with them. Without this input, deterioration can continue unnoticed, names and taxonomy become outdated and specimens become increasingly difficult to find and access. Attempts to revive such dormant and neglected collections require significant resources, and even when staff are available they need to have sufficient knowledge or understanding of the collections to provide more than remedial assistance. Staff and volunteers bring both specific and general skills, knowledge and behaviours to the collections they manage. They facilitate access for borrowers and visitors who in turn contribute their own knowledge to improving the collections. Together they drive a dynamic, organised and growing set of information on the natural world. They are a key element of strategic decision-making processes. Two of the most basic strategic questions are "What is my goal?" and then "Do I or my team have the appropriate skills to achieve that goal?"

Staff and volunteers are your most important and valuable resource, this is especially true in collections management where both skilful hands and creative heads are needed and where staff usually stay in posts much longer than those in research projects. A skilled and creative team working in an atmosphere of mutual trust and respect will achieve a lot more than a group of individualists who neither trust each other nor work together. The latter will likely present you as a manager with pain and problems for years. Investing in your team will help you meet challenges and achieve your goals and hopefully enjoy your job.

Throughout this book we will have referred to a variety of roles and job titles and there are a confusing number of these varying from country to country and institute to institute; there are curators, collection managers, collections leaders, conservators, digitising technicians and many more, and then the managers that oversee all of these; heads of collections, principal curators, directors of collections, heads of the herbarium and more.

Whatever the name, the role is the most relevant to strategy. Since the 1990s roles have changed and new roles have developed. The concept of a collection manager itself is a relatively new development compared to times when these tasks were wrapped up in junior research posts or the responsibility of technical staff. In smaller institutions it has long been the case that staff have roles with multiple responsibilities, from collections management to education and exhibits. On the other hand, the data revolution has created a need for staff with specialist knowledge of systems to focus on managing digital collections (see Chapter 7). Furthermore those who had limited areas of responsibility now have much larger remits; the invertebrate zoology collections manager is also the coordinator of integrated pest management, or the natural history curator in a local museum is the project manager for a move of all types of collections to a new site. A new type of collection manager is evolving with much wider responsibility and competency to fulfil these broader roles.

In all organisations, staff resources are essential to achieving objectives and the underlying principles of people management apply as much in a collections-based organisation as in a business. There is a copious amount of literature available on managing and developing human resources and then a smaller number of sources on management in museums. Rather than repeat the general wisdom available on managing staff, we are conveying our personal perspectives and some of the tools that have been developed to help managers make the most of staff resources, specifically methodologies, for gathering data on the skills, knowledge and behaviours available or lacking in the workforce to support strategic decisions and assist individuals' personal development and mobility. We begin by examining the ways in which organisations with natural science collections arrange their staff, for example into departments/themes, and how this can influence collections goals. We then focus on individual staff and explore a methodology for gathering data on the skills, knowledge and behaviours available and then identifying and addressing gaps to support strategic decisions. Finally, we look at how that wealth of knowledge and skill can be captured and retained rather than lost when staff leave or move to new roles.

Questions

- How do organisations structure their workforce and how can this affect management of collections?
- How can I identify the skills, knowledge and behaviours (competencies) available in my team?
- How can I match competencies to collections objectives and address any that are weak or missing?
- How can I capture and retain knowledge and experience of staff for the future?

How do organisations structure their workforce and how can this affect management of collections?

Structure is essential to staff management in organisations, private or public, large or small. It is essential for chains of responsibility, staff appraisal and development and distribution of tasks to achieve objectives. It is closely linked to governance (see Chapter 4). Collections organisations arrange their staff in different ways depending on factors such as mission, tradition, politics, personalities, size of organisation, type of collection and staff development needs. Individuals within these departments, themes and divisions bring a range of skills and qualifications that vary from professionally- or vocationally-trained researchers, with parts of their jobs devoted to collections management, to full-time collection managers/technicians with extensive training, practical experience and knowledge of collections management. In organisations with a mission focussing mainly on public engagement, structure might be more allied to exhibition and education needs.

We will explore the strategic role of staff structures through examples of organisations with natural science collections and consider any advantages or risks associated with them. Traditionally, larger natural history museums divided their staff into departments according to taxonomic group such as "Department of Entomology" or "Department of Vertebrate Palaeontology". Staff in herbaria might be organised according to responsibility for particular plant groups or regional divisions such as "Tropical Department". In smaller and mixed discipline organisations they might relate to broad disciplines such as "Department of Natural History".

Over the last 30 years there has been a trend in organisations with active research programmes to distribute staff among Departments of Research or Collections Management. In other organisations collections management is carried out by technical staff, sometimes with minimal qualifications, and supervised by those with a research background or specialist knowledge.

In our experience, many organisations around the world follow two main models and variations on them. Although these are derived from larger museums with research functions, in smaller and/or multi-disciplinary museums the research function can for example be replaced by exhibitions and education programmes, which often fall within a collections manager's or curator's remit. We have observed that the advantages and disadvantages associated with these models relate mainly to the varying priorities and drivers of the organisations and the tendency to stick with familiar long-established structures.

We have summarised the main models in diagrammatic form and these show different levels of integration, i.e. the degree to which responsibility for collections and other key functions are distributed amongst staff. There is also a third "hybrid" model where cross-collections roles such as registration, conservation services and biorepository services are located in a centralised collections facilities department.

The two main models are simplified in Figures 5.1 and 5.2 to show the basic principles of each. The light shaded areas indicates the percentage of

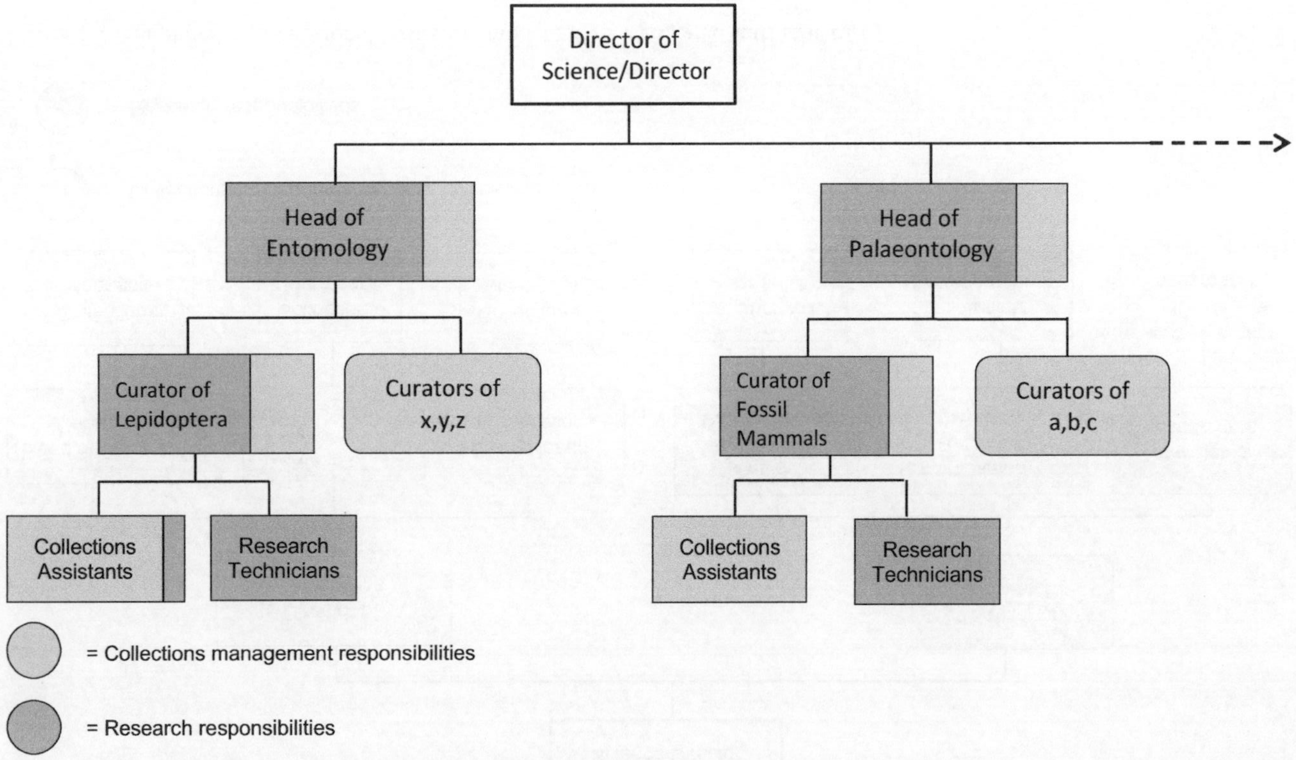

Figure 5.1 Simplified organisational structure: model "A" – integrated staff structure

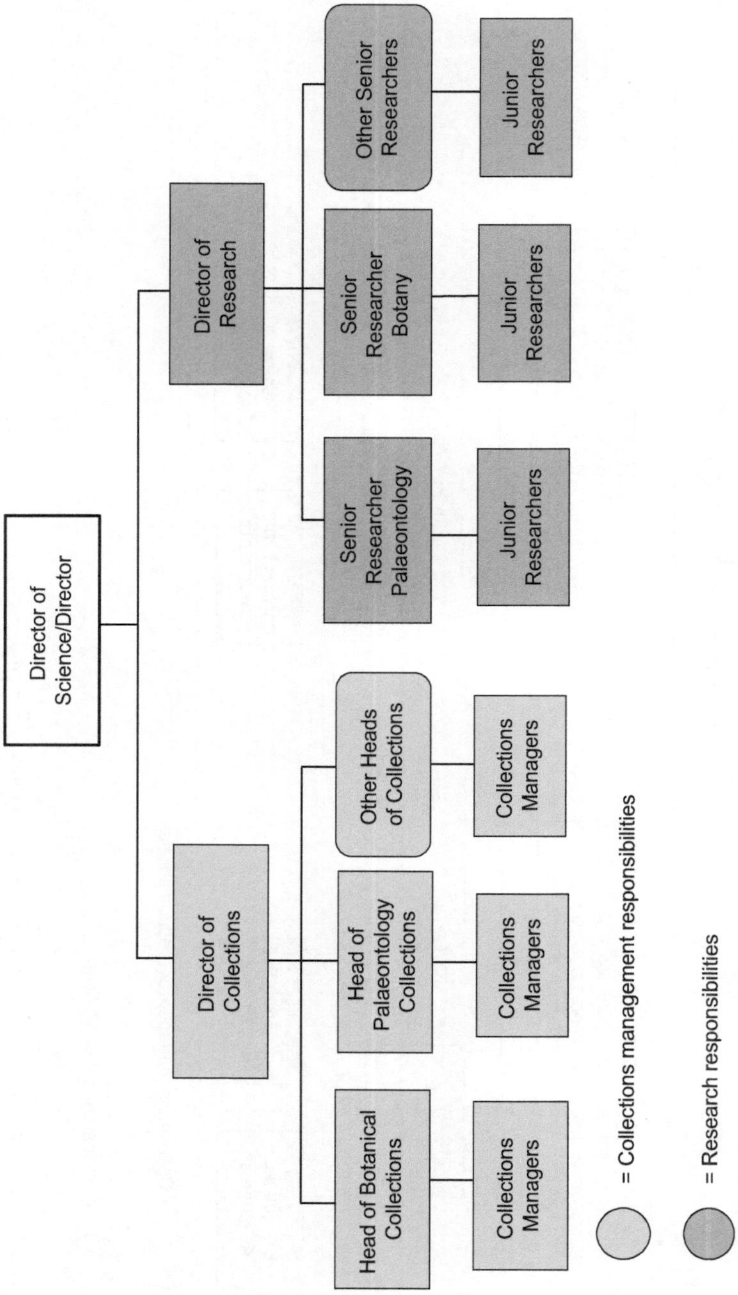

Figure 5.2 Simplified organisational structure: model "B" – segregated staff structure

staff members' time devoted in some way to managing collections and the dark shaded areas indicate research activity. For simplicity we have left out time spent on other activities such as administration, corporate activities, etc.

In model A (Figure 5.1), decision-making responsibilities for both collections management and research are shared among staff with research backgrounds. Sometimes, but not exclusively, more junior scientific staff have a larger proportion of their work devoted to collections management, although basic tasks such as re-spiriting collections are often assigned to technical staff with minimal qualifications.

In model B (Figure 5.2), collections management is entirely the responsibility of specialised collections staff organised in a separate "Department of Collections" with a "Head or Director of Collections" reporting directly to a Director of Science or General Director. Staff in the research section have the vast majority or all of their time allocated to research with some exceptions.

There are many variants of these basic models, for example some larger organisations have separated collections and research functions as in model B, but only within departments led by research leaders.

Strengths and weaknesses of the different models

Staff in model A organisations or similar may feel that they are closely involved with and contribute to the management and development of the collections, avoiding an "us versus them" situation. On the other hand, if career advancement is primarily through high-impact scientific publication there is a risk that staff will focus more heavily upon research and the collections will suffer through less and inconsistent time and resources being devoted to their management and care.

Model B organisations point to the benefits of professional development of their collections staff, sharing of best practice and direct representation of collections at senior level. There is the risk that impediments to communication between departments can lead to research and other staff lacking the opportunity to contribute their expert knowledge and experience, for example to decisions on acquisitions and collections development.

Exercise

Look at model A (Figure 5.1) and model B (Figure 5.2) and consider the following:

- Which of these does your organisation most closely follow?
- List any strengths and weaknesses of your own structure.
- What might you do, if anything, to address any weaknesses?

How can staff structure facilitate good collections management?

In addition to the traditional role of providing chains of management and responsibility, we believe that staff structure can contribute to best practice in management and care of collections. To do this they should:

- enable sharing of expertise and knowledge of best practice across the organisation;
- enable standard approaches to collections management where appropriate;
- ensure cost effectiveness in purchase of materials, collections furniture, etc.;
- ensure that performance in collections management and care is recognised and rewarded, as is research performance, and allow career progression for collections management staff;
- allow input of research expertise to collections decisions where appropriate;
- encourage joint research and development projects based on collections management issues, e.g. conservation science applied to processes of deterioration;
- avoid "selfishness", e.g. research or collections staff refusing access to the groups of organisms which they are studying;
- allocate authority and responsibility at consistent levels across the organization so that staff compensation and opportunity is logical and fair;
- facilitate career progression and pathways for staff specialising in collections management.

What if the structure is not working?

Sometimes structures will not meet some or all of the criteria listed above and may not be ideal for meeting collections goals. This may be for reasons of legal factors, past agreements and more. Significant changes may not be possible or affordable in terms of time and money. Such changes will not be within the control of staff below director level or higher.

There are actions however that can be taken independently of staff structure and these are listed below. Before introducing any of these measures it is wise to carefully assess the situation! What actually is the problem and where is it? Are there inefficient procedures, incompetent staff members (wrong staff in the wrong place), or maybe conflict between individual staff members? Make sure that your staff are involved in the problem-solving process and that you have in-depth talks and interviews with your staff to find out more about background and possible solutions. Then take action – for example:

- Have a senior staff member with experience and engagement with collections management to represent the interests of collections at a senior executive level within the organisation.
- Introduce cross-organisational collections groups with representation from internal stakeholders in the collections (e.g. conservators, collections managers, researchers, exhibitions and education staff, facilities managers). Ideally give this group clear purpose and status through responsibility for delivering defined cross-organisation collections improvement projects.
- Ensure that all staff with collections responsibilities have these clearly written into their job description with targets and performance expectations, regardless of their main area of expertise. For example a researcher who also has collections responsibility would be expected to perform well in both these roles.
- Give staff who have the potential for exceptional competency as strategic managers of collections the opportunity for reward or progression in the departmental structure.
- Ensure that all staff have clearly written performance goals and that the staff discuss these goals at the beginning of each review cycle with their work leader.
- Ensure that team leaders are trained in overseeing staff work and that they are ready and committed to developing their staff.
- Develop processes to encourage and allow mediation to solve conflicts between staff members and prevent this having a disruptive effect on the organization.

Staff structure is not something to be changed lightly or regularly. Doing so can be disruptive and expensive because it can require revising job descriptions, creating or updating significant work plans, or responding to staff concerns about their pay and their future job prospects. Any risks posed to collections by staff structure can be minimised through proper considerations of the advantages and disadvantages of different models, and if a structure cannot be changed, finding other ways to minimise risk.

How can I identify the skills, knowledge and behaviours (competencies) available in my team?

Whatever the staff structure or size of your institution you will need a spectrum of knowledge and abilities to achieve your goals. Knowing what strengths staff and volunteers can bring to the task and if there are gaps and how might they be filled to achieve goals is a key element of strategic planning. Regardless of structure, people bring a heterogeneous range of abilities and knowledge to their role in the organisation. They may also have strengths that are unknown to team leaders and managers. For example experience brought from previous jobs or learned during their development in their role. People are arguably the organisation's greatest strength but ensuring that the attributes needed are in place or are known to managers requires information gathering.

There are a number of tools and methods available to gather this information, many of which have been used in business for many years. For example, skills gap analysis is often used to identify whether the skills of the staff match those needed to achieve organisational goals. These analyses are usually carried out to meet a specific need, drawing upon information garnered from staff using focus groups and surveys. These skills analyses are often carried out by, or with the help of, human resources departments. Another more permanent and consistent means of recording information on skills, knowledge and behaviours that has found use in recent years are competency frameworks, and we will focus on these in this chapter.

Competency frameworks are a comprehensive, structured variation on gap analyses used extensively across a wide range of organisations from health care to human resources. Competencies can be regarded as the skills, knowledge and behaviours required to do your job effectively or to achieve an objective or goal. Frameworks capture competencies in a structured way, enabling individuals to identify their training and development needs and for managers to identify the competencies available to them to achieve strategic goals. There are frameworks designed specifically for museums, for example, compiled by the UK Collections Trust for use by staff in a wide range of museums and galleries (Dawson, 2015). It is a high-level framework which can be used as a starting point to define, plan and manage the skills needed by employees of a museum.

Some frameworks can be more specific to the roles and tasks that are expected in a museum environment. For example, the *Dictionnaire de Compétence en Gestions des Collections* (Société des musées du Québec, 2013), developed by a consortium of museums in Quebec, provides a common language to clarify roles of museum professionals. It is a list of the tasks that are needed to effectively manage a collection rather than a list of competencies, although one can be derived from the other.

Since 2015 there has been a framework available designed specifically for natural science collections. The European Union Collections Competency Framework (EUColComp, http://eucolcomp.myspecies.info/) is based on a framework used successfully for staff development and progression in the Natural History Museum, London, since 2008 and can be used to define the skills, knowledge and experience that a curator/collection manager/conservator needs to manage a scientific collection at any particular level, from beginners to experienced professionals and leaders.

The need for such a comprehensive framework specific to natural science was identified by the transnational projects SYNTHESYS (www.synthesys.info/ref) and the European Distributed Institute of Taxonomy (EDIT).[1] A joint working group identified notable variation in the level of training and experience of collections staff across the European Community as a potential barrier to present and future biodiversity research. The solution identified was to develop a set of competencies specifically designed for the complex needs of natural science collections that could be applied in any institution regardless of language and governance. The aim was to create an integrated infrastructure for biodiversity research that crossed national boundaries, and a crucial part of this would be the ability of staff to move easily from one institution to another, i.e. staff mobility. The project was eventually addressed by a third European Union project, the European Union Collections Competency Framework (EUColComp). This universal set of competencies is available in seven major European languages and allows individuals to easily and unambiguously describe their competencies as an aid to their development and increased job mobility within Europe. It is a matrix of competencies arranged in four levels of progression, from early-career to established professionals, and covers the wide range of activities that those with responsibilities for natural science collections are expected to carry out competently.

Figure 5.3 shows an example of one of the competencies relating to integrated pest management. The complete framework can be found at http://EUColComp.myspecies.info/.

	Level A Early-stage career, training level	*Level B Established professional*	*Level C Mid-career, professional, manager*	*Level D Senior manager, expert*
1.4.iii. Integrated pest management	1403A	1403B	1403C	1403D
Collections management core competency	Is aware of importance of all IPM and quarantine procedures.	Understands and applies procedures for managing pest infestations without supervision.	Advises on management of pest-related issues.	Reviews and approves IPM policy and procedures for the Institution.
Conservation core competency (to level B)	Able to follow IPM and quarantine procedures with minimal supervision.	Carries out regular pest monitoring programmes.	Reports at departmental level on IPM and coordinates monitoring programme.	Ensures IPM policy is implemented and adhered to.

Figure 5.3 Sample competency from the EuColComp Framework

Note that in this institutional example all collections staff are required to have attained, or will soon attain, level B in this competency.

Source: EUColComp was funded through the European Commission, Leonardo da Vinci Transfer of Innovation Programme.

There are three important points to note regarding these frameworks. Firstly, no staff member would be expected to possess all of these competencies or be required to do so. A dedicated conservator requires different but overlapping competencies to, for example, a collection manager. Some competencies however can be defined as core and an established collection manager would be expected to be competent in these, at least to the level of "established practitioner". Secondly, the level of progression does not necessarily link to pay grade or position, although there is potential for them to be used in that way. The different levels of competency required in an organisation will vary depending on factors such as size and structure. For example, a director of a small organisation might be expected to have a similar set of level "C" and "D" competencies to a "Senior Collections Manager" responsible for a specific collection in a large national organisation. Thirdly, although developed within the EU the framework can, with respect to local legal requirements, be applied internationally.

When an individual records the competencies they currently possess they create a personal competency profile. Figure 5.4 shows a simplified profile for an early-career collections manager, showing some of the 92 competencies in the framework.

In order to understand how these frameworks can be used, and how appropriate they might be to your own organisation or team, we have provided some simple exercises.

Exercise

In this exercise we aim to familiarise you with frameworks and their strengths and weaknesses using EUColComp as an example. A much shorter, simplified version of the framework is provided in Appendix 5.1. You can copy this to record your own levels of competency and consider its value to individuals and managers.

Part A

Read through each cell of the example matrix and then if you think that is a competency that you possess colour it in with marker pen or annotate it in some way that makes it stand out. Assuming that you possess some but not all of the competencies present you should have a profile not necessarily the same as, but similar to Figure 5.4.

Part B

Having experimented with the sample framework:

- Was it easy to complete and if not, why not?
- How could you provide evidence that you have that competency?

Collections management. Conservation and Care				
Competency	**Level A** **Early-stage career, training level**	**Level B** **Established Professional**	**Level C** **Mid-Career, Professional, Manager**	**Level D** **Senior Manager, Expert**
Collections Development Strategy: Development	Provides data e.g. from collections surveys to inform collection development strategy, under direction.	Identifies areas suitable for development such as gaps or limited coverage.	Implements collection development strategy with specialist input from team.	Develops collection development strategy for the relevant Area of Responsibility.
Collections Development Strategy: Assessment	Becomes familiar with collection within designated area of responsibility.	Demonstrates an understanding of the strengths and weaknesses of the collection, e.g. what people want/need; keeping up to date; knowing the gaps etc.	Establishes criteria to help determine strengths and weaknesses of all collections within area of responsibility.	Establishes criteria for assessment of strengths and weaknesses collections care and holdings.
Collections: Risk Management	Handles material in the correct manner.	Understands risk management as applied to collections.	Is up to date on the latest information and techniques on mitigating the effects of deterioration.	Maintains the Risk Management Plan for the Department.
Integrated Pest Management	Is aware of importance of all IPM and quarantine procedures.	Understands and applies procedures for managing pest infestations without supervision.	Advises on management of pest related issues.	Reviews and approves IPM policy and procedures for the Museum.

Figure 5.4 A simplified EUColComp competency profile for an early-career collections manager, showing they have acquired the double-lined competencies

Source: EUColComp was funded through the European Commission, Leonardo da Vinci Transfer of Innovation Programme.

How can I match competencies to collections objectives and address any that are missing?

Competencies can be used to identify what skills you *have* as an individual but they can also be used to identify the skills that you *need* to achieve team objectives. The next exercise allows you to compare what competencies you have and what you need.

Exercise

Looking at the competencies in Appendix 5.1, which of these do you think you need to carry out for either your own job, a collections management or conservation job with which you are familiar, or maybe a position that you would like to reach in the future? Using a copy of the sample framework from Appendix 5.1, mark these on the copy you used in part A of the previous exercise, but use a different colour or some other means that allows both to be visible. You should end up with a framework that resembles Figure 5.5.

Now compare the two profiles.

- Do the two sets match? Are there any competencies that the job requires that you don't have?
- How might you gain those competencies?
- Was this exercise easy to complete and if not, why not?

Figure 5.5 provides a simple example of matching the competencies you have to those that you need and identifying any gaps.

Matching competencies to organisational needs

One of the advantages of these more detailed frameworks is that they can be used by managers, team leaders and others to identify the competencies needed to achieve their aims and then to see how that matches to the competencies available in their teams. For example, a museum or group of museums discover that among their collections are human remains collected many years ago. They are confronted with the sensitive ethical and legal challenges associated with these collections but do any of their staff possess the necessary skills and experience to manage these collections? Completing and collating the framework for the team will show who, if any, has these skills and whether that is sufficient for the collection's needs. The options can then be assessed: do you train existing staff, recruit people with those competencies or maybe both?

	Collections management. Conservation and Care				
	Competency	**Level A** **Early-stage career, training level**	**Level B** **Established Professional**	**Level C** **Mid-Career, Professional, Manager**	**Level D** **Senior Manager, Expert**
1	**Collections Development Strategy: Development**	Provides data e.g. from collections surveys to inform collection development strategy, under direction.	Identifies areas suitable for development such as gaps or limited coverage. **?**	Implements collection development strategy with specialist input from team.	Develops collection development strategy for the relevant Area of Responsibility.
2	**Collections Development Strategy: Assessment**	Becomes familiar with collection within designated area of responsibility.	Demonstrates an understanding of the strengths and weaknesses of the collection, e.g. what people want/need; keeping up to date; knowing the gaps etc.	Establishes criteria to help determine strengths and weaknesses of all collections within area of responsibility.	Establishes criteria for assessment of strengths and weaknesses collections care and holdings.
3	**Collections: Risk Management**	Handles material in the correct manner.	Understands risk management as applied to collections. **?**	Is up to date on the latest information and techniques on mitigating the effects of deterioration.	Maintains the Risk Management Plan for the Department.
4	**Integrated Pest Management**	Is aware of importance of all IPM and quarantine procedures.	Understands and applies procedures for managing pest infestations without supervision. **?**	Advises on management of pest related issues. **?**	Reviews and approves IPM policy and procedures for the Museum.

Figure 5.5 Sample EUColComp competency profile for an early-career collections manager (double-lined) and the profile that is required for their job (shaded). The question marks indicate competency gaps and development needs

Source: EUColComp was funded through the European Commission, Leonardo da Vinci Transfer of Innovation Programme.

These exercises can be done on paper copies, but by entering the framework into a spreadsheet and applying formulae, a number of other analyses can be carried out, for example how many staff in your team possess any particular competency. Also, profiles can be matched automatically and areas in need of development highlighted. Using a spreadsheet and applying formulae also makes it possible to collate all the profiles in a team and identify how many staff have a particular competency. Figure 5.6 shows an example of a "heat map" profile for competencies in digital curation. The darkest squares indicate where there are few or no staff with that competency at a particular level.

This is a simple example and one could ask why the manager could not just ask the staff if they have any experience in this area? The advantage of the framework is in its consistency of language and level of progression. This is also only one area of many collections activities. When all the tasks and objectives of both individual and team are considered, the framework presents a complete and consistent picture.

Are all competencies equal and are they all needed?

Not all competencies are equal. One person would never be expected, or ever be able, to possess all these competencies at all levels. However, there may be some that are essential for any person working with collections on a regular basis as part of their job. For example, any collection manager in the EU would be expected to be competent at least to EUColComp level B in collections risk assessment. These are termed "core competencies" and they may vary depending on role. Those for a conservator might differ from those for a collection manager or digitising technician, but they will overlap.

	Levels of progression			
Competency	A	B	C	D
3.i. Data Entry	27	26	18	5
3.ii. Query and Reporting	27	26	17	7
3.iii. Data Validation and Enhancement	27	25	17	4
3.iv. Mass Digitisation: Projects	23	19	11	7
3.v. Mass Digitisation: Digital Imaging	25	23	14	5
3.vi. Workflows and Process Management	26	23	18	6
3.vii. Manage Collections Data : Quality Control and Assurance	27	26	13	3
3.viii. Manage Collections Data: Policies	26	21	9	6
3.ix. Database Management: Technical Interface	24	19	13	5
3.x. Database Management: Advanced Database Management	1	1	2	0
3.xi. Collections Management System: Development	26	18	9	2
3.xii. Collections Management System: Registry management	26	16	7	1
3.xiii. Collections Management System: Training and Support	27	22	10	1

Figure 5.6 Part of a spreadsheet combining competency frameworks of 27 individuals in a large museum. Numbers in cells represent staff with a particular competency. This "heat map" shows where competencies are missing (darkest shading)

Source: EUColComp was funded through the European Commission, Leonardo da Vinci Transfer of Innovation Programme.

Examples of application of the framework

The ability to combine frameworks in spreadsheets has proved a useful tool in assisting organisations to identify developmental needs. Here are two examples: one from a large national museum and the other from a consortium of smaller organisations.

The EUColComp framework has been used to identify the numbers of staff with particular competencies and how well that matches the strategic needs of the organisation. This can then be used to identify what training or recruitment is needed to achieve the organisational goals.

Example: Using a framework to identify staffing gaps

The National Museum of Natural History of the Smithsonian Institution (NMNH) wanted to gather accurate and current data about their existing staffing counts and areas and levels of expertise to establish an optimal level of well-trained staff across their collections. In 2016, NMNH used the EUColComp framework to assess existing competencies and identify development needs. The framework was used to analyse both the competency needs of management and the competencies held by staff. Learning opportunities, or other options such as "buy-in" or redeployment, can then be identified to address gaps.

By entering their data into a spreadsheet version of the framework each staff member could see a graphic representation of their competencies and where any gaps might lie. Each staff member indicated which competencies he or she had or didn't have with a "Y" or an "N". The spreadsheet highlights all cells where individual have responded "yes" in green, giving a profile graphic which the staff member can use to identify personal development and refer to in order to measure any change through time.

The spreadsheet was used to count all the staff holding a particular competency. Figure 5.6 shows a "heat map" for the "Digital" tab of the framework.

The exercise showed that within the group participating, staff have a distribution of competencies expected from an institution of the size and caliber of NMNH, i.e. a pyramidal distribution with the clear majority of staff having a wide range of competence at the A and B levels of progression, with fewer at the more senior levels C and D. However some gaps in competence were identified both within functions and within departments. Suggestions were then made for development to address these gaps. For example, gaps in the management of human remains could be addressed by internal training/shadowing by a colleague with the missing competence, studying the human remains policy in detail, carrying out an audit of the collections and, finally, by discussing these with their colleagues and supervisor.

Example: Using competencies to share skills between museums

A regional consortium of museums holding natural science collections in the UK identified a dramatic decrease of specialist expertise in managing and conserving their collections. This was partly due to staff reductions but also a trend towards employing more generalist staff. The framework was used to gather data on skills and competencies needed to manage the collections across all eight organisations and then identify what the gaps where and how these could be addressed. Heat maps similar to Figure 5.6 were generated and areas could be identified where they could share competencies across the consortium rather than the clearly impossible task of having all skills in each organisation.

Using competencies to build jobs and teams

Job descriptions are an essential means to articulate what an organisation needs from its employees. They show where a job fits within the team or organisation and defines the boundaries of responsibility between one staff member and another. There are a number of ways of building job descriptions or revising existing ones to match the current and future needs of the organisation. There are many sources of information available on the generalities of job analysis. Competency frameworks such as EUColComp and the Quebec Task Dictionary can provide a set of unique and clearly described elements that can be used to "build a job". Every cell in the EUColComp framework can be seen as one of 368 (92 x 4 matrix) uniquely numbered building bricks. For example competency 1.4 at level A in Figure 5.3 is coded 1403A. The advantage of this coding is that cell 1403A can be translated into any language and moved around but always refers to the same competency. The building bricks can be put together to construct job specifications and job profiles created using a clear unambiguous language that is tailored to the needs, legal requirements and structures particular to any organisation. This aspect of competency frameworks was applied effectively in the reorganisation of scientific staff at the Museum für Naturkunde, Berlin (MfN) (see example below).

Example: Using competencies to build job descriptions

The Museum für Naturkunde, Berlin, Germany evolved from a fully integrated model A museum. Staff in collection management reflected a patchwork of a historically grown staff structure. Job description and salary level was rather a matter of coincidence and good connections between line management (from collection management staff, via curator and head of department to the director) than of actual needs and expertise. There was no career pathway. Both led to non-transparency and frustration. The EUColComp framework built the basis for the development of a specific Museum für Naturkunde competency framework. In the process, building bricks (cells) of the original framework were arranged in a way that meets the legal and structural requirements of the organisation. These job descriptions were designed as building blocks that enable a career progression from level A ("Sammlungspfleger") to level C/D ("Senior Collection Manager"). By taking out basic level competencies and adding new qualities from one step to the next, job descriptions from A to D are linked to each other and build on each other. The implementation of the framework is now creating transparency and the chance for career progression in collection management. Both can be seen as milestones. Furthermore, the design of the Museum für Naturkunde competency framework is connected to the development of new overarching staff structures in collections management that allows more teamwork and efficiency of processes.

How can we address any missing skills, knowledge or behaviours?

Like most organisations, museums and herbaria are rarely without some form of training and development plan for staff. Training is usually available in generic areas such as problem solving, planning and courses on more specific topics such as integrated pest management. Once gaps in competency have been identified using the frameworks, managers and staff can address these. This ability to assess the training and development needs of staff is one of the most valuable aspects of frameworks.

Exercise

Looking at the competency profile that you created earlier, consider the following:

- How might you gain any missing competencies: a training course, an exchange visit, experience, mentoring?
- What might be most appropriate or available?
- Do some competencies fall into groups that could be addressed through one learning resource? For example could several competencies be gained through a general course in strategic planning?
- What development options are available either in your organisation or beyond?

The framework can be used as a starting point for identifying learning resources. Many competencies, particularly at higher levels, are specific applications of more general competence. For instance, to successfully reorganise a collection requires planning skills which can be gained through formal training and experience. Others, such as understanding the strengths and weaknesses of a particular collection, can be gained through experience or by working with more experienced staff and gaining a competency through "knowledge exchange" (see section below on succession planning).

Strengths and weaknesses of frameworks

Competency frameworks can be criticised as being too rigid and for not achieving what a good manager could find out by "knowing their staff". However, they are not intended to replace managerial skill and managers' understanding of their teams, but instead to provide transparency and a consistent vocabulary to assist discussion on development and building teams to meet strategic goals. If the framework is comprehensive and well-structured then it should be possible to modify it to address new competency needs in collections work. There are specific strengths and weaknesses of the EUColComp framework and many of these were identified over years of use of the original framework at the Natural History Museum, London. The EUColComp framework itself was tested and reviewed by staff and managers over several months in eight European natural science collections organisations. The potential benefits and weaknesses identified are listed below. How does this compare with your framework test in the exercises above?

Strengths

- Achieving a shared organisational purpose and vision.
- Improved long term planning for museums, governing bodies and learning providers enables museums to identify skills gaps.
- Improved personal career planning and Continuing Personal Development (CPD).
- Improved performance management.
- Improved understanding of roles and responsibilities and better team working.
- More focussed/appropriate recruitment.
- Recognition – framework highlights staff abilities and gives them a sense of recognition even if pay increases or promotion are not available.
- Transparency – managers and colleagues can see what all team members have to offer and in recruitment this could be a safeguard against discrimination (also see weaknesses below).
- Understanding the resources available to managers in theirs and other's teams.

- A tool to make the case for funding more posts.
- Can provide graphical evidence of skill gaps to potential funders or senior management.
- Improved mobility of staff within and between countries.

Weaknesses and areas for improvement

- Size and complexity: drawing up profiles for individual staff using 92 competencies and four levels can be time consuming and daunting.
- Can be seen as a potential threat: if not introduced in a participatory way staff might perceive the framework as a way to identify potential staff reduction.
- Possible clashes with legal frameworks.
- If not managed carefully, staff may have unrealistic expectations – merely having the competency does not guarantee promotion for example.
- Staff need to be clear that to possess a competency they must demonstrate performance with evidence.
- Not all staff would be happy with others viewing their competency profiles.

Addressing weaknesses

Frameworks can be very helpful in staff development and mobility but can also be seen as a threat. Whether developing a new framework to suit your own team or using an existing framework, the importance of involving relevant staff and management at all stages cannot be overstated. The acceptance and use of the Natural History Museum, London framework is, to great extent, due to its development by a team of managers, collections staff, human resources representatives and trades unions.

Succession planning: How can I capture and retain knowledge and experience of staff for the future?

A competency framework will give you a clear idea of what skills and knowledge you have in your team. It will also starkly illustrate what you are losing when a staff member leaves. The retention of knowledge is an ever-present challenge that has always been there and is an inherent weakness of organisations such as museums. Here we explore this issue and look to describe some measures that you can employ to minimise the effects of loss.

The act of making collections has a long history in cultures and locations around the world. Over the centuries that collections have been assembled, we have learned a lot about physical aspects of care: preservation, arrangement in high-quality containers, environmental quality and appropriate buildings, handling, agents of deterioration and so on. Manuals have been written with standard terms and recommendations for information management. Collections management has come into its own as a profession. Yet, we still lose information with the loss of staff and this is a frequently voiced justification for hiring new staff before the retirements of an earlier generation of staff. Administrators ask for demonstrations or non-hypothetical examples of what information is lost if a collections staff position is vacant. In general, the answer is sobering. Staff that have extended experience managing a collection over many years develop a deep understanding of its history, special qualities and potential. They know its history of use, its problems and especially what work has not been done that would otherwise ensure its safety, preservation and future.

The reality of collections work in institutions of all sizes is that some work goes undone, just as some maintenance is deferred. How does a manager respond to this problem and build systems that prepare for staff vacancies, especially those that arise with short notice? How does an institution overcome a long tradition of oral transmission of collections "lore" to managed documentation of work systems, products and statuses?

Here is a list of methods that have been used to ensure that valuable information held in multiple documents and in the memory of staff, volunteers and research visitors can be retained to enhance and preserve collections.

To preserve and share collections information you can:

- record and transcribe oral histories from departing staff and insist that any collections project carried out has a written summary and that key documents are updated and accessible to all when the project ends;
- centralise archiving of critical information using a standard organisation scheme about how the collections are managed so they become available to all;
- have centralised, digital systems for recording information about the collections, for example using wiki portals or similar vehicles for finding information on a particular topic;
- keep checklists of key data, preserved in digital form;
- create "fact books", a register with all relevant data and information about a specific collection: size, users, highlights, impact, accessibility that can be regularly updated;
- gather data in summary form rather than specific form;
- set up organisational mechanisms, such as cross-organisation groups, to promote a sharing culture (see section on staff structure above);
- try to ensure overlapping posts – if possible start someone in a new post before the previous holder leaves.

You will never be able to capture all the competencies and group knowledge that your staff or your fellow team members possess but hopefully the suggestions on this list will mitigate against this loss and avoid endlessly "inventing wheels".

Summary

Staff are the most valuable resource in your organisation or project. An understanding of the staff resources available to achieve your goals is essential to strategic planning, whether of a natural science collection or a large businesses enterprise. In this chapter, we have identified some of the key questions that a manager might ask about the staff resources available to them, whether paid or voluntary, few or many. We began by examining the structures across which the staff resource is distributed. Ideally structure may be aligned to the mission and objectives of the institution and benefit the long-term preservation, development and accessibility of the collections.

Organisations arrange their staff in many ways, and in museums and herbaria this falls mainly into two or three models depending on degrees of specialisation in collections care and management compared to research and other functions. Although this is not something that you may be able to change directly, understanding the merits of different structures may help you to contribute and advocate in any institutional senior discussions on restructuring to minimise any risks. If structure cannot be changed you can consider some of the other ways that we have listed to ensure communication and interaction between collection staff and researchers, educators and other sectors. Ultimately nearly all models can work well if you have the right competencies in the right place and allow for communication and interaction, transparency and staff development.

The second and perhaps most important question we consider is the staff themselves, arguably your greatest asset in achieving your strategic goals. We evaluated the potential of competency frameworks as tools to aid personal development and for managers to develop their staff resources to achieve strategic aims. We used the EUColComp framework, developed and actively used by the authors, as an example of a framework that is tailored to natural science collections. It provides a means to map progression and gives a transparent view of an individual's capabilities and development needs. It paints a picture of the resources available to managers to achieve their goals and highlights any gaps to be addressed through development and training. The changing demands on staff managing natural science collections has created a need to develop skills and knowledge in new areas that would have not been dreamt about by most collections workers 40 years ago. Skills in advocacy, developing virtual collections, dealing with increasingly byzantine legislation, trends towards greater job mobility and expectations make open discussion based on documented information more important than ever. We believe this is a useful tool that, along with regular discussion, helps you to know and respect your staff or project team and their capabilities, often beyond the obvious.

Inevitably, the competencies and accrued institutional knowledge of your staff will be lost when they leave or move elsewhere in your organisation for example. Frameworks can help you think about exactly what competencies you are losing and how they might be replaced. Knowledge accrued over tens of years may also be lost and we listed some mechanisms to mitigate against this.

What we have proposed and demonstrated above may seem an unobtainable ideal when governance and lack of funding prevent you from changing structures, promoting staff with high-level capabilities or recruiting new staff. But hopefully the tools and suggestions we have provided can help you work within these limits to build stronger teams to achieve your goals whatever the scale. Finally we would ask you to consider the value of collaboration with other institutions as a means to share training costs, exchange knowledge and fill competency gaps in your own area of responsibility.

Whichever tools or methods you choose to employ or perhaps develop for yourself we cannot overestimate the value of involving your staff or fellow team members in that process.

Note

1 EDIT (European Distributed Institute of Taxonomy) was funded by the European Union and ran from 2005 to 2010 and aimed to reduce the fragmentation in European taxonomy through increased cooperation and integration between the larger taxonomic research institutions.

References

Dawson, V., 2015, *Developing a Collections Management Framework*. Collections Trust. https://collectionstrust.org.uk/resource/developing-a-collections-management-framework/ [Accessed 05/11/19]

Société des musées du Québec, 2013, *Dictionnaire de compétences – Gestion des collections*. www.musees.qc.ca/activites-publications/smq/dictionnaire-de-competences-gestion-des-collections [Accessed 05/11/19]

Further reading

Arts Council England, 2016, *Character Matters*. www.artscouncil.org.uk/sites/default/files/download-file/Character_Matters_UK_Museum_Workforce_full_report.pdf [Accessed 05/11/19]

Genoways, H.H., and Ireland, L.M., 2003, *Museum Administration – An Introduction*. Walnut Creek, CA: Altimira Press.

Lord, G.D., and Lord, B., 2009, *The Manual of Museum Management*. Walnut Creek, CA: Altamira Press.

Williams, S.L., and Hawks, C.A. (eds.), 2006. *Museum Studies: Perspectives & Innovations*. Washington, DC: Society for the Preservation of Natural History Collections.

APPENDIX 5.1: SIMPLIFIED COMPETENCY FRAMEWORK

This is a simplified version of the EUColComp Framework (http://eucolcomp.myspecies.info/). You can use this to plot your own competency profile and match it to the requirements of your job, or maybe a job that you aspire to. Think about what is missing and how you might address those gaps through training or experience for example.

EUColComp was funded through the European Commission, Leonardo da Vinci Transfer of Innovation Programme. The materials hereby freely made available for use have been created by the Natural History Museum, London in conjunction with its international partners in the EUColComp project and we would kindly ask that this is acknowledged in any re-use, re-purposing or dissemination of these materials. This framework reflects the views only of the authors. The Commission cannot be held responsible for any use which may be made of the information contained therein.

Table 5.1 Simplified version of the EUColComp Framework

		Level A *Early-stage career* *Training level*	*Level B* *Established professional*	*Level C* *Mid-career* *Professional Manager*	*Level D* *Senior manager* *Expert*
	Collections management – conservation and care				
1	**Collections development strategy: development**	Provides data, e.g. from collections surveys to inform collection development strategy, under direction.	Identifies areas suitable for development such as gaps or limited coverage.	Implements collection development strategy with specialist input from team.	Develops collection development strategy for the relevant area of responsibility.
2	**Collections development strategy: assessment**	Becomes familiar with collection within designated area of responsibility.	Demonstrates an understanding of the strengths and weaknesses of the collection, e.g. what people want/need; keeping up to date; knowing the gaps, etc.	Establishes criteria to help determine strengths and weaknesses of all collections within area of responsibility.	Establishes criteria for assessment of strengths and weaknesses of collections care and holdings.
3	**Collections: risk management**	Handles material in the correct manner.	Understands risk management as applied to collections.	Is up to date on the latest information and techniques on mitigating the effects of deterioration.	Maintains the Risk Management Plan for the department.
4	**Integrated pest management**	Is aware of importance of all IPM and quarantine procedures.	Understands and applies procedures for managing pest infestations without supervision.	Advises on management of pest-related issues.	Reviews and approves IPM policy and procedures for the museum.

5	**Disaster planning: general**	Understands their role in collections rescue and recovery, i.e. emergency planning, and is familiar with disaster planning arrangements for the area of responsibility.	Manages day-to-day business to reduce risks likely to initiate a disaster.	Is a key member of the department/institution's salvage team and provides advice for rescue of all collections within their area of responsibility (i.e. they are the main emergency contact for that area).	Instigates, develops and manages emergency planning, i.e. business continuity, disaster, and collections salvage plans for area of responsibility.
6	**Collections care and conservation: collection condition surveys**	Able to follow survey protocols and contribute effectively to condition surveys.	Plans a condition survey following recognised survey methods/protocols agreed by team or team management.	Audits and provides overview of surveys and evaluations of collection status, e.g. comparing state of collections to recognised benchmarks/standards.	Sets standards, based on condition survey data, and, where relevant, international best practice.
	Digital curation				
7	**Data entry**	Can use the agreed system/software for managing collections accurately and quickly to incorporate data.	Is sufficiently conversant with the operation of any collections management systems to carry out routine tasks such as searches, loan lists, etc. accurately.	Manages digital curation effort for team and/or area of responsibility.	Chairs departmental and/or museum committees for coordination and prioritisation of digitisation effort.
8	**Query and reporting**	Able to interrogate collections management systems/databases, etc. to provide information.	Uses collections management systems/databases to perform complex queries about holdings.	Develops report formats for user needs. Uses collections management systems/databases to assess collection status and gaps in area of responsibility and to respond to enquiries from external groups.	Can fully manipulate collections management systems/databases to provide reports to satisfy senior level requests and key performance indicators, e.g. number of loans sent in a year.

(*continued*)

Table 5.1 Cont.

		Level A *Early-stage career* *Training level*	*Level B* *Established professional*	*Level C* *Mid-career* *Professional Manager*	*Level D* *Senior manager* *Expert*
9	**Mass digitisation: projects**	Provides curatorial support for mass digitisation projects, e.g. providing specimens.	Suggests projects for mass digitisation and supports development of large import/update files.	Manages data entry projects in excess of 100,000 records.	Designs, develops and manages mass digitisation projects.
	Scholarship				
10	**Systematic and taxonomic principles**	Learns the basic rules of nomenclature, i.e. understands type concept.	Understands the rules of nomenclature as they relate to their discipline.	Expert understanding of the use of rules of nomenclature and codes and can relate the importance of the collections to scientific debates, discussion and enquiries.	
11	**Scholarship: general**	Develops a working knowledge of an assigned area or collection.	Has a working knowledge of an assigned collection and relevant literature.	Has a comprehensive knowledge of all collections within area of responsibility and relevant literature.	Has exceptional knowledge of the area of responsibility and relevant literature, e.g. highly familiar with collectors represented, knows collectors handwriting, can relate collections to collections in world context, etc.

12	**Scholarship: identification skills**	Is able to identify relevant material to an appropriate level under direction.	Identifies material without supervision and to a high level of accuracy. Has an understanding of where to source the relevant literature, e.g. Zoological Record.	Provides identification skills to support all division/section/team collections based activity.	Uses expertise and experience to advise others.
	Exhibitions and outreach				
13	**Exhibitions: content and development**	Provides information and specimens to be used in exhibitions as instructed.	Provides informed suggestions, copy and specimens for exhibitions.	Part of the development/design team for an exhibition.	Provides expertise and leadership in the development of a large exhibition.
14	**Exhibitions: loans**	Is familiar with conservation issues for loan material. Can pack objects for loan under supervision.	Reviews facility reports and other associated documents for loan of material relevant to their section.	Acts for the department or institution to ensure criteria for display of specimens are met by borrowing departments and institutions.	Carries out a range of registry duties for specimen loan for exhibitions. Coordinates and manages area of responsibility/exhibition loans.
	General management				
15	**Resource management**	Uses available resources in an effective way, avoiding waste and inefficiency.	Understands the financial implications of uses of resources.	Understands procurement processes, i.e. rules governing purchase of goods and services within the institution.	Puts together clear, focussed and justified bids for funding within institution and to external funding bodies.

6 Acquisition and deaccession

Introduction

Over the centuries, the development of natural science collections has been variable, from deliberate to quite haphazard, driven by curiosity and discovery, reflecting, amongst other things, colonial expansion and fashion. In the late 20th century collections institutions, especially larger ones, gradually focussed on acquisition strategies to reflect their research needs. To gain a better understanding of the natural world other than by random collecting, but also driven by a changing international world order in which it was no longer acceptable to send collectors to other countries to collect whatever they could, museums simply had to have a plan; a focus on what and where to collect, now in close cooperation with and gaining permission from local partners in countries of origin. The principle of not collecting more but collecting smarter – quantity versus quality – became a key element of strategic management. Museums have also increasingly confronted the controversial and arguably essential role of deaccession in the development of collections. Challenges of space and resources to manage all collections in one institution pose questions on what might be better housed and used elsewhere. Rarely would collections be actively destroyed, strategic decisions on proactive deaccession being rare if at all allowed. This chapter aims to provoke thoughts on the strategic and ethical application of acquisition and deaccession of objects or collections. Reactive deaccessions are where institutions have been forced to close or through changing priorities and increasing financial restraints have resulted in valuable collections becoming "orphaned", i.e. unmanaged, inaccessible and at risk due to damage through neglect or disposal as trash. We provide thoughts to assist organisations needing to dispose of collections and those who might acquire these collections to do this in a way that ensures their long-term survival and accessibility. Exchanges of collections are actually a combination of acquisition and disposal. Collections can also be improved and made more useful by the addition of new information, for example by identification of specimens, addition of more accurate geographical data and by adding data from new "modern" research such as DNA and isotope studies.

Worldwide, institutional collections grow as more and more objects are added and many new museums, for instance in China, appear. There is a challenging task ahead of us to further promote the scientific and cultural value as well as the use of natural science collections. Keep in mind that deaccession is not that different from acquisition, it is just the other side of the coin. Both are supposed to improve the quality of a collection, helping to meet collection profiles and hence mission statements.

Questions

- Why should we acquire and deaccession objects and collections?
- How can collections be acquired?
- What criteria do you have to take into account to acquire or deaccession objects or collections?
- How can you deaccession objects or collections responsibly?
- What about the ethics of collecting?
- Are there alternatives if deaccession is not an option?
- How should I communicate deaccessioning to staff, collectors, donors and the public?

Why should we acquire and deaccession objects and collections?

Institutional mission statements determine collection profiles (see Chapter 3). But as mission statements change over time, collection profiles change. Collections are dynamic. But do you consider and treat your collection as being dynamic? You probably do as you add objects on a regular basis which support your research programmes and exhibits. But what do you do with objects or collections which no longer support your mission? Have you ever considered disposing, or even actively disposed, of material?

The purpose of acquisition and deaccession is to optimise collections and have a cost-benefit effect. With the ICOM Code of Ethics for Museums as a precondition, acquisition and deaccession have to be carried out within the restrictions of cultural, national and internal laws and regulations such as the Convention on Biological Diversity (CBD). It means that some institutions, depending on ownership of the collection and country of residence, have more freedom to acquire and especially deaccession than others. The options will therefore differ between institutions and countries.

Who takes the decisions in your institution? In the case of deaccessioning it will generally be taken at the highest level. However, in natural science collections it is common practice as part of daily routine that decisions on deaccessions are made at a relatively low hierarchical level. For example, a collection manager of a botanical collection might decide on exchanging duplicate material, as in botany, flowers or leaves from a single plant, bush or tree are effectively the same collection event or, in other words, identical objects. Distribution of such duplicates between institutions diminishes the risk of loss and increases the opportunity to study the material, reducing the need to travel and loan. In the case of particularly valuable or sensitive objects however, decisions should be made at a directors' or even ministerial level. Given the scale of natural science collections this is a pragmatic approach with the majority of the decisions taken at a lower level, not putting the burden on those higher up in the hierarchy. The same would be true for other kinds of collections dealing with large numbers of closely related objects as long as the criteria, for instance regarding the amount and kind of (irreversible) damage or decay which makes disposal acceptable, are clear and formulated and approved at higher levels. But do collections staff or (in large institutions) researchers initiate proposals for deaccessioning in your museum? Do they consider acquisition and deaccession important enough to spend time on? Do they know and embrace the collection profile as discussed in the collection management or development plan or should they be triggered top down to do so? It might be helpful to set out guidance on who has authority to do what. In their Collections Management Policy, the Smithsonian National Museum of Natural History has a scale based upon resource needs, dollar value, significance, etc. such that some acquisitions or deaccessions can be made up to a certain level by a curator, while others must go through an approval process at higher or even the highest level. Authority and responsibility for acquisition and deaccession should be part of and detailed in competency frameworks to avoid any confusion or unauthorised activity (see Chapters 4 and 5).

In the 1960s, a large-scale move was made between the Natural History Museum (NHM), London and The Royal Botanic Gardens, Kew. The driver then was rationalisation and avoidance of duplication of research focus. Bryophytes, algae and lichens went from Kew to be amalgamated with the existing collections at the NHM. In return non-lichenised fungi moved to Kew. These moves were part of the "Morton Plan" to delineate research responsibilities for the two largest botanical collections in the UK. Discussions on a broader national plan continued into the 1990s.

Acquisition and deaccession do not necessarily have to be the result of an institutional vision as captured in a collection profile. It might also be imposed for other reasons. Limited storage space and labour, time and money might force you to choose between quality versus quantity. Less *can* be more! If resources cannot be spent on collections that are considered to add little to the value of your collection, then you might consider finding a more appropriate place for such collections and focus on more important assets to your institutional profile. For example, a museum might decide to dispose of European art because it can only focus on traditional arts of its region given the loss of sustainable budget, in which case its mission should be adjusted relative to the available resources and/or expertise.

Exercise

Consider the following:

- What is the mission of your institution and what does this mean for the collection under your care?
- What parts of the collection fit the mission best and why?
- What parts can be developed and which objects or collections do not fit the mission?
- Can these be considered for deaccession?
- If not, why not if they do not seem to fit the mission of your institution?

How can collections be acquired?

If you are reading this as a researcher or collection manager in natural history you will be familiar with acquisition of specimens by collecting in the field, but there are of course a number of ways to acquire. These methods might not be applicable for each institution as some are related to or depend on research focus or budget. We review these briefly below.

Expeditions

In natural science, archaeology and ethnography scheduled expeditions might bring in considerable numbers of objects which fit the museum's mission and hence its collection profile, more now than in the past when "as much as possible" was collected randomly. Historical collections dating back to colonial times are still the geographic backbone of scientific focus, which determines collection strategy and international cooperation of many museums. However, most institutions now have a much better focus when collecting or are restricted by national and international regulations.

Donations and bequests

In the case of donations, philanthropists and amateur biologists and geologists are an important source as they frequently donate their collections to a museum, hoping for the continued existence of their collections on which they spent so much time, money and love. They may also donate or deposit as a condition of a grant or collecting agreement. All sorts of collections are out there waiting for us. A survey by Naturalis Biodiversity Center among amateur entomologists in the Netherlands in 2017 revealed the presence of at least 90 private collections of good scientific quality, totalling 1.5 million specimens, which would be donated in the coming 20 or so years. Depending on the collection policy, which might however change over time, a good deal of these 90 collections might be considered valuable additions. Careful planning and early support, e.g. by donating storage units or help with digitisation, might increase your chances to acquire these collections more easily and spread the workload of its incorporation and accessibility over time. As in many countries, donations are tax deductible, so you have something to offer to the donor in return, apart from the fact that his or her private collection will be well cared for and become publicly accessible and available for the future.

Sales and auctions

The mechanism of acquisition of objects through sales or auctions is a well-known principle and requires little attention here. In the framework of natural science collections, sales and auctions are regular and legal means to acquire mainly fossils, amber, minerals, stones and gems but also large insect and shell collections for example. However, there are ethical aspects of purchasing objects commercially that require special care and attention. Is it clear and officially documented how and from where these objects originated? Were they legally acquired and exported? The case of the dinosaur remains illegally exported from Mongolia, auctioned in the USA, and later seized by law enforcement officers and returned to Mongolia is a clear warning to be extremely careful and always double check the provenance and legal status (www.bbc.com/news/world-asia-22431009).

Transfer or exchange

Transferring or exchanging objects or collections with other institutions to optimise your own collection as part of your museum mission and collection profile is worth consideration and benefits both. In natural science, the number of samples taken during expeditions might be too large to keep them all as vouchers. It is common practice that scientists or collection managers donate or exchange the surplus, or even entire taxonomic groups considered as "by-catch", with other institutions where experts are working on these groups. In botany, so-called duplicates such as stems with leaves and flowers from the same plant, or in mycology parts of the same lichen, are frequently exchanged, as is the case with insects when large numbers from the same species from a single locality have been collected. They are exchanged for specimens of other species or donated to compensate for time spent by staff of other institutions on their identification if the expertise is not available in house. In larger institutions, scientists or collection managers often decide on this. But are they indeed the right person to do so without senior approval? In natural science collections, it is considered common practice and generally accepted. In smaller institutions with generalist rather than specialist staff, decision taking might better be a team process and approved at a higher management level to put responsibility where it should be.

Orphaned collections

Too often, institutions cease to exist or no longer want responsibility for collections, putting entire collections at risk. We are all aware of collections that were not taken care of and in time fell apart, ending up as public exhibits where documentation became separated from the main collection rendering them useless for scientific studies, in students' "private collections", in the hands of dealers or in the trash. National or international networks might help to save these orphaned collections and become your means of acquisition of objects which fit your collections profile, saving valuable objects which will remain available for use in the public domain.

Example: An orphaned collection

During the economic crisis of the early 21st century, a regional natural science museum in the Netherlands with 125,000 objects had to close its doors. It was a member of the "Stichting Nederlandse Natuurhistorische Collecties" (SNNC), a foundation of 12 Dutch natural science museums. The collection, which was about to become orphaned overnight, was divided among the members after a consensus was reached quickly and, after some discussion, unanimously. In the event of differences in opinion it was up to the owner of the collection to have the final vote. The collection was split up into 18 lots. Regional collections were returned to museums in the region of origin. Collections and objects of national and/or international importance went to the national collection. Also, the collection on display found a new home, attracted a new audience and was made available to other SNNC members afterwards. Leftovers not falling within either of these categories were donated to visitor and educational centres. In cases where no one was interested in acquiring them, the objects were destroyed. This procedure differed slightly from what is described in the *Guidelines for Deaccessioning Museum Objects* (LAMO, Museumvereniging, 2016) (see below); in this particular case there was no reason to include art, antiquity and/or anthropology museums in the procedure because of the purely natural history aspects of the objects. Partnership, communication and common sense turned the inevitable and regrettable closure of one institution at least into a successful and happy end for the orphaned collections.

What criteria do you have to take into account to acquire or deaccession objects or collections?

Objects under your care support the mission of your institute because of their scientific, cultural or historical value. Or because you consider them suitable for exhibition, education or even strategic reasons. Strategic reasons can be, for example, acquisition of specimens for exchange purposes, possible investment for the future or policy driven based on the request of your institution's financier or owner of the collection if you care for instance for a government-owned collection.

When objects or collections fit one or more of the above criteria, contain voucher or type specimens or have been digitised, you have reasons to acquire them or keep them as part of an institutional collection. When objects do *not* fit any of the above criteria there seems to be no reason to keep or acquire them. If objects are already part of your collection and do not fit any of the above criteria they can be considered for deaccession by applying the same criteria as for acquisition. You can remove items without any scientific information from your research collection and give them a second life by putting them in an educational or public collection. If safe from a health and safety perspective, this is a way of continuing use while managing the refinement of your scientific collection. Although this might not create immediate space, which could have been your reason for deaccession, it does help lower management costs as these items are allowed to be "loved to death" by the public, after which they can be discarded. Aspects of health and safety, ethics and legality always have to be taken into account, as for example with radioactive minerals or illegally acquired goods.

How can you deaccession objects or collections acceptably?

Adding to your collection is what we all seem to find normal. Ask around in your institution, ask your colleagues and the vast majority considers acquisition, not deaccession, the way forward. But when discussing this issue more thoroughly, mind-sets might change.

Exercise

Organise a discussion in your department with collection managers and/or researchers about deaccession. Ask both groups for their arguments, both for and against, and list these. The next step is to prioritise these arguments with input from both sides and see whether the discussion changes points of view of the group as a whole.

In many countries, objects or even entire collections can be exchanged with, donated or transferred to other museums. They will remain available as museum objects although in a new location. This means that institutions with objects or collections that do *not* fit their profile could thus deaccession these to institutions where they *do* fit the profile. It turns disposed collections into acquired collections which *will* be properly managed, stored, studied, exhibited and/or used otherwise: a win-win situation.

Deaccession of objects by selling or destroying them is not allowed in most institutions. But even if it is, this should be the last option. Responsible decision taking is most important here. There are reasons to decide to sell or destroy objects. Imagine a collection of birds or mammals that have been exhibited for many years, or even decades, and which originated not from the wild but from captivity and hence have less or no scientific value. The objects may have lost their colour or became damaged or for any other reasons have become unsuitable for future exhibitions in your museum. If they do not serve – for your organisation – any further and future scientific, cultural or educational purpose, these objects might be deaccessioned to other museums or might even be ready for destruction (see Disposal Toolkit, below). The answer will differ between institutions, their mission and the scale on which they work. When objects or collections are sold or destroyed, it can even be considered a win-win situation. From a cost-benefit point of view, curatorial investment and space will no longer be spent on collections without direct relevance to the owner. Their destination and reasons for disposal need to be added to the collections database, even after its physical destruction to create transparency and clarity to former owners and future users of their whereabouts or what has happened to the objects. And be aware that it is unethical for staff to take personal possession of items disposed by the museum, for any purpose, even if they are taken out of the trash bin.

The *Disposal Toolkit – Guidelines for Museums* compiled by the Museums Association (2019) is a very helpful UK manual which includes a disposal flowchart and chapters entitled "Why dispose", Methods of disposal", "Recording the process", "Dealing with problems" and "Communicating the process", which help you through the process (www.museumsassociation.org/collections/disposal-toolkit). Similarly, Spectrum, the UK collection management standard, including references therein, is a very helpful tool (https://collectionstrust.org.uk/spectrum/).

The so-called *Leidraad Afstoten Museale Objecten* (LAMO; Guidelines for Deaccessioning Museum Objects, Museumvereniging 2016) is the official procedure Dutch national collections have to follow when deaccessioning objects. The LAMO serves three main objectives:

1 Protection of heritage of national importance, so that it is preserved for Dutch society.

2 Support for the collection management of museums, as an instrument to promote collection mobility and as protection against omissions and mistakes in that process.
3 Promoting social trust in the museum sector, as guardian of the national heritage.

It provides flowcharts and text to help you carefully plan to complete the entire process of deaccessioning and disposal, which includes, among others, notification of donors prior to disposal to maintain good relationships.

As many objects, especially with older collections, have been treated with toxins used in preservation, such as arsenic and mercury as well as biocides and pesticides, you should be aware of the dangers involved from a health and safety perspective. This is true not only when you donate your objects to other institutions but also when you destroy them. In natural sciences, many of the objects will have to be treated as chemical waste when discarded. This involves extra measures and costs.

What about the ethics of collecting?

Shooting and collecting 26 orangutans for scientific reasons, as Alfred Russel Wallace did in Borneo in the mid-19th century, or killing the last 44 great auks in 1844 for commercial deals with museums, is fortunately no longer common practice as, since 1975, it is restricted by international rules and regulations such as the Convention on International Trade in Endangered Species (CITES, since 1975) and common sense. Collecting more common, less appealing vertebrates as part of research programmes has also become more difficult to get accepted. But what about insects and molluscs, for instance? Until recently no one asked questions about collecting and killing them but that is changing as well. A paper by Alan Hart in *The Biologist* in 2018 (Hart, 2018) reported on a citizen science project in the UK in which people were asked to assist in collecting wasps – wasps which many of us loathe and readily kill if we get the opportunity. However, killing wasps to study their distribution and abundance attracted considerable criticism in the national press, in which it was labelled a "bizarre conservation project" and "condemned by wildlife experts". What do you think? Has the way we do research and acquire collections changed in recent decades and if so, how and why? If (even?) killing or collecting wasps is debatable, then what is not debatable? Consider how you might enhance your collections through acquisition, what might the constraints be and why?

Indeed much has changed. Borders between European countries might have disappeared for EU citizens allowing them free crossing but they still remain barriers for transfer of objects which fall under national and/or international regulations. Bureaucracy has increased tremendously, with extra costs involved for both staffing time and permits. The Nagoya Protocol (see Chapter 4), an agreement through which benefits from genetic resources are shared between the holder of the collection and countries of origin of the genetic source, has increased the administrative burden for research collections not only prior to collecting expeditions, but also afterwards when material is studied, sent out on loan, analysed, published or in some cases made into a profit. The Nagoya Protocol therefore has major consequences for natural science collections and studies based on such collections. An extensive manual to help steer collections through the legislation, compiled as a joint effort by the natural science community, expresses the complexity of it (https://cetaf.org/news/cetaf-code-conduct-abs-features-peer-reviewed-article-research-limitations-abs-context).

Social media does influence our attitude towards collecting and killing live specimens, especially when vertebrates or, for instance, large or beautiful butterflies are involved. This understandably dissuades institutions dependent on public support and killing and collecting vertebrates has significantly diminished. Alternatively, expeditions now frequently focus on collecting samples such as blood, feathers, hairs or tissue for DNA studies, releasing the specimen alive and hence losing your voucher. This might be more readily accepted by the public and countries of origin than collecting and killing. However, it does not make bureaucracy any less, as legislation such as Nagoya or CITES do not differentiate between small samples or complete specimens. Moreover, future studies will be hampered if voucher specimens are no longer available, an issue of serious concern.

Will natural science reach a stage where living creatures will no longer be collected because of ethical restrictions imposed by ourselves or by the outside world? If invasive sampling (blood) or extraction of a leg of an insect for DNA analysis when the animal itself is not killed is out of the question, perhaps we have to work with what we have without expanding further. In botany and geology, public opinion will not play a significant role, but the rules and regulations, as recently formulated in the Nagoya Protocol, certainly do. Collecting ethnographic objects and taking them out of their cultural context as we frequently did in the past is no longer considered uniformly ethical and in some cases has led to the return, by deaccession, of such objects to the countries of origin, the original owners or a museum in that country. This is, however, still an issue of much debate causing tensions between museums and governments north and south of the equator.

Maybe not an ethical issue, but certainly a new regulation is the new European law on privacy, the so-called General Data Protection Regulation (GDPR) which came into effect on May 25th, 2018. In the field of scientific collections, names of collectors are linked to objects and included in databases. Holding records of collectors' activities affects their privacy as it makes the whereabouts and activities of those individuals public. He or she needs to give approval to link and/or publish his or her name in conjunction with the objects. Time will tell whether this issue resolves itself or becomes unworkable. When GDPR, which is mandatory, came into effect it seemed that natural science collections were not ready for it or had not even considered the consequences.

Although the benefits of acquisition and deaccession for institutions are obvious, national and international regulations have made the process much more complicated than in earlier days. It is, however, part of our mission to answer questions on biodiversity and evolution and consequently to develop our collections as a global research infrastructure. International cooperation between natural science institutions is of utmost importance if we are to avoid duplication of effort, use organisms more efficiently, use historical collections retrospectively, build consensus with conservation organisations and communicate effectively with the outside world in this era of expanding social communication. Without these it will become increasingly difficult, if not impossible, to study biodiversity and biodiversity change now and in the future. Time series, for instance, are and will remain important assets to monitor our environment and the changes taking place in this era of global change. It will require continuous sampling of zoological, botanical and geological specimens worldwide. Ethical consideration, good arguments and answers to sensitive questions will become more important than ever before.

Example: Acceptable or not? A case study

In 2016, an undescribed species of bird was discovered in a national park on a tiny island. It was absent from any natural science collection. Because of the small size of the island, the population was, and must always have been, very small. The authorities of the national park agreed to allow the collection of a single specimen to be described as a new species, as this will put it, and hence the national park, "on the map" and increase the conservation status of both the bird and the national park. The killing of a single specimen will, in their opinion, attract international attention and help protect flora and fauna on the island. From a biological point of view, the death of a single individual should or will not harm a healthy but small population as the place of the individual will be taken by a young one. However, the ethical question here is whether the killing of an individual bird, removing it from a very small population, is acceptable for reasons mentioned.

Are there alternatives if deaccession is not an option?

Whether due to the restraints of governance, for example, you are unable to improve the efficiency of your collections through deaccession there may be other options.

Collection storage and management as a business model

Outsourcing the storage and/or management of collections is an option worth considering as an alternative to the improvement of old or the construction of new facilities. Storage of collections in facilities of other institutions as a business model can (often quickly) improve the conditions under which collections are housed, be a cheaper option for the owner than any alternative and generate income for the receiver. It differs from deaccession in the fact that these collections do not change ownership and will continue to be managed via curation agreements. Bringing together even very different kinds of collections from various institutions into shared collection storage facilities can reduce costs as it may lead to more efficiency in curation and access to these collections.

Updating collections data

Collections with limited value do not necessarily have to be considered for deaccession. The value of objects and collections can be improved and made more useful for multiple purposes by the addition of new material or new information. In natural science collections, identification of specimens, results of analysis from DNA studies, addition of accurate geographical data by geo-referencing and biographical and historical information will increase the value of objects. Linking field notebooks to collections raises the value of the objects as they often contain a plethora of information on exact location, date and circumstances of collecting. Drawings may provide information on colouration of animals and plants before they lost their brightness due to death and then aging of the specimen. This requires careful management, the help of specialists, amateurs or even the general public by making such information available through digital exposure, open access, citizen science, etc. The "Making Sense of Illustrated Handwritten Archives" project brings together organizations specialized in advanced computer and data science, artificial intelligence and cognitive engineering and natural history. They cooperate to digitise, transcribe and annotate the notes and notebooks of the so-called "Natuurkundige Commissie", which collected animals and plants in the Dutch East Indies (now Indonesia) between 1820 and 1850. This invaluable project enriches all these objects increasing their scientific value tremendously (https://sites.google.com/naturalis.nl/makingsenseproject/homepage).

How should I communicate deaccessioning to staff, collectors, donors and the public?

Communication is an important part of deaccession, as indicated in the Disposal Toolkit and LAMO. It should have both an internal and external focus as it is not only museum staff who will raise questions about the why and how of a disposal. Also, the original collector or donor, if at all possible or realistic, as well as the public (or, if you will, the tax payer) need to be informed and deserve an explanation. The public needs to know that deaccession is a legitimate, prudent part of collections management. Otherwise there can be claims of wrong doing or open disagreements with deaccessions. For example the loss of objects in which public money has been invested for many years or concerns about popular, favourite specimens from exhibitions, even if they have been destroyed for good reasons following the appropriate protocols. Deaccessioning, as shown above, does not necessarily include "destruction" which is likely to be the degree of disposal which needs most explanation. Other forms of disposal will be more easily acceptable to the public. Internal communication, documentation and transparency are equally important to keeping your staff up to date and making them aware of your decisions and process.

Summary

Institutions managing collections should ask themselves on a regular basis whether collections under their care still fit their mission statement and collection profile, i.e. the written expression of strategic thinking and strategic habits. Acquisition is generally considered an ongoing concern which, however, should be a responsible and carefully considered process and decision based upon your acquisition policy. For example, is it wise to accept collections now which will later be judged as of no value to your organisation as they do not fit a future collections policy and will be disposed of by your successors? Museum management should similarly consider whether some objects or collections are not better off elsewhere. Not only in other institutions, but also if deaccessioning would mean permanent and definite disposal by means of destruction. If, in the end, an institution deaccessions parts of its collection to meet its collection profile, and another institute will manage, update, study or exhibit that collection, then deaccessioning has been successful.

Increased international legislation, economic development and social media have greatly changed the way with which scientific expeditions are mounted and specimens are collected, exported and studied. It is expected that the growth of natural science collections with "fresh" specimens will quickly diminish and that we will have to focus on what we already have. Use and re-use, improved methodologies and new questions will enable us to make optimal use over and over again of the collections which we so carefully manage. Whether our future scientific and societal questions can be answered remains a question in itself.

References

Hart, A., 2018, Inside the Killing Jar. *The Biologist*, 65 (2): 26–29.

Museums Association, 2019, *Disposal Toolkit – Guidelines for Museums*. London: Museums Association. *www.museumsassociation.org/collections/disposal-toolkit* [Accessed 05/08/19]

Museumvereniging, 2016, *LAMO 2016*. www.museumvereniging.nl/lamo-leidraad-afstoten-museale-objecten [Accessed 05/08/19]

Further reading

Ekosaari, M., Jantunen, S., and Paaskoski, L., 2015, *A Checklist For Museum Collections Management Policy*. Museum 2015 Project and National Board of Antiquities; The National Board of Antiquities' guidelines and instructions 9. www.museovirasto.fi/uploads/Arkisto-ja-kokoelmapalvelut/Julkaisut/museum-collections-management-policy.pdf/ [Accessed 05/08/19]

Morgan, P.J., and National Museum of Wales, 1986, *A National Plan for Systematic Collections: Proceedings of a Conference Held at the National Museum of Wales, Cardiff, 6–9 July 1982, Cardiff*. National Museum of Wales.

Museums & Galleries NSW., 2011, Deaccession and Disposal In Small Museums Fact Sheet. https://mgnsw.org.au/sector/resources/online-resources/collection-management/deaccessioning-and-disposal/ [Accessed 05/08/19]

Smithsonian Institution, 2005, Acquisition and Disposal of Collections. In: *Concern at the Core: Managing Smithsonian Collections* (pp. 139–202). Washington, DC: Smithsonian Institution. www.si.edu/content/opanda/docs/Rpts2005/05.04.ConcernAtTheCore.Disposal.pdf [Accessed 05/08/19]

University of Alaska, Museum of the North. Deaccessioning and Disposal. www.uaf.edu/museum/collections/ethno/policies/deaccessioning/ [Accessed 05/08/19]

7 Virtual collections

> A digital collection is the twin of the physical collection but unlike the physical has the potential to be aggregated, linked and used as part of one very big worldwide collection which exists apart from institutions.
>
> (Vincent Smith, Natural History Museum, London)

Introduction

The demand for access to digital data and images is growing rapidly. While scientists, both academic and amateur, have known how to find their way into our collections for a long time, the general public, however, was mostly dependent on our exhibits to see and learn about our collections. But now the general public has become an important client for studying or simply enjoying the beauty and value of parts of our collections to which they never had access before. In addition, through rapidly booming citizen science projects, we ask the public to help us answer important biodiversity questions. For that reason, but not only, it is our desire and obligation as museum managers to make collections under our care accessible to everyone, with as few limitations as possible. Hence major institutions are digitising their collections on a massive scale, with increasing speed and using newly developed fit-for-purpose techniques. Our desire to expose our objects through the virtual world is offering new opportunities to present collections to an ever-growing audience more quickly and in different ways than ever before. The rapid development of specialised software and emerging technologies such as pattern recognition has helped us to achieve this. Simple databases no longer satisfy user communities' information needs and they demand digital images. Effective digitisation to meet these needs requires complex decision making on methods, standards and legal considerations. Museums are building upon the experience of others and now apply international standards to present images of the highest quality. By using additional tools, virtual objects can show much more detail than the physical object will ever reveal to the naked eye. Digitisation has also become a major competitor for funding, whether it is from within budgets or external, including crowd-sourcing or crowd-funding. Involvement of "the crowd", whoever or wherever they are, is a powerful way to add content, and hence additional value, to our collections.

Despite the many advantages of digitisation, there are multiple pitfalls. Virtual collections do not, as many believe and even suggest, replace physical collections (see Chapter 2). One of the fundamental arguments is that virtual objects do not allow analysis of the physical material from which it is made or its genetic structure. They do not allow for any future analysis of characteristics using techniques that we are not currently aware of. One cannot "take a look" inside a photo as one can do inside a physical object. Virtual collections are not a replacement but an addition to the corresponding physical collections.

Digital techniques are changing rapidly over time. For this reason this chapter is not about techniques, hardware and software, but will focus on *strategic* management, drawing upon experience and examining factors to take into account when embarking on virtual collections projects. More technical and up-to-date information can be found in specialised journals, on websites and directly from institutions that are in the process of digitisation.

Questions

- Why should you digitise your collections?
- How do you create a digitisation strategy and plan?
- What should be digitised, how do you prioritise and who should do it?
- Which standards, software and hardware should you use?
- How much does it cost?
- What digital initiatives exist around the globe?
- What are the risks?
- What is the future for digitisation?

Why should you digitise your collections?

The advantages of digitisation are multiple. Information linked to the objects as well as the objects themselves, in fact entire collections, can be consulted on a global scale by a single push of a button. In terms of reduced risk, there is often no longer the need for the external and internal user to visit a museum or collection and physically handle fragile and valuable objects. For the collection manager there are none of the tasks associated with managing visitors, such as screening and assisting visitors or checking the collection after a visit. The institution will benefit as there will be no disturbance of the climatic conditions in the collection storage facilities because of the presence of multiple visitors. Much – but not all – collection management and research can be done away from where the objects are stored. Access will partly, or even largely, shift from physical to virtual collections. This however, asks for the end-users' confidence in the correctness and completeness of the data transcribed from the labels or captured as an image. Under pressure of time and the desire to be cost-efficient, errors are easily made, especially when for example 19th century handwriting has to be transcribed or when transcription is done by individuals not familiar with the collections and/or poor knowledge of geography. Validation of these data by your own collection managers or researchers or collaborators is of utmost importance. It needs to be emphasised that if digitised data are not trusted and users detect errors or omissions, it will not be used by the scientific community, and if it is used, errors might lead to incorrect reproduction and false conclusions in scientific publications. As managers of scientific collections, therefore, we have a responsibility to ensure the correctness of the digitisation process and hence the balance between efficiency, cost and output. Virtual collections are not only for the general public as a virtual "exhibit", but have to serve scientists, historians and other researchers, including amateurs, around the world and hence have to meet the highest standards of accuracy.

Strategic benefits of digitisation include:

- Access – virtual collections give remote, global access to objects and related collection information, such as field notebooks and letters, which are otherwise hidden in archives and museums.
- Linking of object and collection information – by gathering, improving and combining digital information patterns and links can be made visible which otherwise are difficult to discover. Digitisation makes it possible to combine data quickly and easily on various parts of a specimen such as skin, skeleton, measurements, stomach contents and parasites as well as images, drawings and publications. This would otherwise be a laborious task.
- Validation of data – access to digital data allows the global community of scientists and amateur scientists to validate, correct or add data to objects which improves the quality of research. There are, however, serious risks associated with this as errors can be introduced (see below under "risks").
- Validation of identification – high resolution images, for instance, can reveal new information or information otherwise difficult to capture, which can also assist in the identification of similar objects in the collection. Automated wing-pattern recognition allows the identification of bumble bees in drawers of hundreds of objects and inconsistencies such as misidentifications can be resolved (see http://wing-id.naturalis.nl/).
- Retrieval of missing objects – digitisation helps to retrieve objects that were either identified incorrectly (see above) or have been misplaced among other objects, in drawers, cabinets, etc. in the past.
- Access to virtual collections – the effects digitisation will have on the number of visitors to physical collections or requests for physical loans and the consequences to updating information on the original object has not been answered unequivocally yet but is something you will

have to take into account when making strategic decisions. This will be discussed in Chapter 8.

- Unite distributed collections across multiple institutions – aggregate data to be statistically significant, adding scientific rigour, and supporting "Big Data" questions.
- Part of disaster preparedness – images of objects and labels do not replace the original object but document and save at least the status, condition and data. This is important in the case of deterioration processes, such as aging of mounting media in the microscopic slides, which makes objects less visible or blurs details through time. An image documents the status at the time of digitisation and can help future studies of the original status of the object. In case of damage or loss, caused for example by agents of deterioration such as insect pests, theft, vandalism, flooding, fire or war, information about these objects remains retrievable. Think of what happened with historical collections and objects in the wars in Syria and Afghanistan, much of which was destroyed or sold on the black market, and the devastating fire which destroyed millions of objects in the National Museum of Brazil in Rio de Janeiro in 2018. With images, at least a virtual collection remains. Similarly images can act as evidence of origin in the case of stolen artefacts offered for sale.

The new dimension: the history of mass digitisation

Why mass-digitisation? Natural science collections are vast, counting tens of millions of objects in the major international museums. It is simply impossible to digitise them in a one-by-one traditional way as was done in the early days. It is a simple calculation that, when it takes you five minutes – and that is an optimistic estimate – to handle, database and capture an image of an object the traditional way, you will do 480 objects per week or "only" 21,600 objects per year (45 working weeks). A collection of "only" 1 million objects will take you 46 years to complete or 46 staff for the duration of one year. That is no longer an acceptable option for the data-hungry user community and so we had to look for new, alternative business models, such as partnerships with external, more commercial partners, and we needed bigger budgets. This new approach was revolutionary only a decade ago.

Current projects are built on a legacy of digitising initiatives of varying scales and degrees of collaboration developed from the late 1980s. Often individuals would see the benefits of creating databases of specimens of taxonomic groups that they were interested in. Others might use databases as a new and better means to create catalogues of type specimens. A number of important initiatives began in the botanical community, such as barcoding of objects initially to record and track those going on loan and by extension to link label data to databases. Capturing of digital images of type and historical specimens became commonplace. These various data capturing projects used a variety of technologies and database platforms, often varying within institutions. Bringing these disparate datasets together and merging them was a major challenge even before making them accessible via the internet. This process was made much easier by the data standards developed from 1985 by the efforts of Biodiversity Information Standards (TDWG), a group of mainly collections-based biologists. The sharing of experiences with different methodologies and technologies has led to the standardisation of image resolution and database fields often incentivised by large-scale funding such as the Global Plants Initiative which has required common standards and used relatively simple and affordable technologies to capture and make available much-needed images of plant type specimens in the world's herbaria. These factors have led to the mass of specimen data and images that are now available to all.

Large-scale digitisation of natural history collections really started in 2010 at the Muséum National d'Histoire Naturelle (MNHN) in Paris with the 2D scanning of ten million herbarium plants. It was revolutionary at that time. This was followed later that year by a 13 million euro project to digitise, almost at an industrial scale, a targeted 7 million zoological, paleontological, geological and botanical objects in Naturalis Biodiversity Center in Leiden, The Netherlands. On average, excluding money for management and staff, 1.85 euro was available per object. During the five-year project, processes were refined, new techniques developed and manuals written. Cost efficiency was crucial, leading to efficient digitisation of as many objects as possible. These initiatives triggered other institutions who visited Paris and Leiden to discuss, learn, copy and improve their own processes. But what seemed state of the art only five to ten years earlier has now become common practice in many collections worldwide. Outsourcing parts of or even entire processes, as in the digitisation of botany collections, to commercial companies has led to new business models, higher standards, higher output and lower costs. The same will happen over the next decade(s), for example with mass digitisation of insect collections which still set real challenges due to the complexity of label scanning and the fragility of the objects. The Museum für Naturkunde (MfN) in Berlin which, in 2018, was granted about 90 million euro for the digitisation of its entire collection of 30 million objects, might be best prepared to take up this challenge!

How do you create a digitisation strategy and plan?

Effective digitisation and access requires strategic decision making and in-depth discussions, advice on methods, standards and legal issues and benefits. So what is the strategic approach to take, especially by management and staff with less exposure to collections digitisation, such as those from a non-science or non-collections background?

Strategic approaches in a rapidly changing world

Increased access to collection data, standardised data storage according to international standards and linking the virtual with the physical object lead to multiple benefits, including:

- embedding digital management in workflows, leading to quicker and better collection services, lower cost of management and maintenance of physical collections and easier decision making;
- better data quality on websites, facilitating research among a wider audience;
- increasing availability of data of focus groups such as endangered, invasive or harmful species, allowing for rapid assessments;
- the advantage of delivering easily accessible statistics for scientific and management purposes;
- improving the exchange of information compared to the costly and potentially harmful exchange of physical objects;
- more efficient use of storage facilities through better insight into the size, composition and usage of your collection.

The question that is frequently asked by finance officers, funders and others is "What benefit, including generating income, will we see from this expenditure"? This is a difficult question to answer as sometimes benefits may not be realised immediately. In this way, digital collections do not differ from traditional collections. We invest in their preservation and accessibility knowing that we are providing a research facility that may lead to scientific discovery. It is not necessarily linked directly to the work that we and others currently do on them. On-demand digitisation, however, generating data for third possibly commercial parties, can generate income and encourage cooperation. It will help third parties instantly without them having to spend much time and effort in gathering these data from scratch or in any other way.

Exercise

A series of benefits of digital collections have been given above. What other benefits can you add, especially and specifically for your own collections?

The digitisation process

The digitisation process can be divided into three stages:

1 *Mobilisation*: connecting the physical to the digital. Mobilisation is the process of transforming data on labels for example, to digital form

and includes equipment, methods of transcribing data such as crowdsourcing, optical character recognition (OCR) and enhancing data with georeferencing for example.

2. *Systems*: once the information is in electronic form there is a need for improving data quality and putting it into context such as climate research and resolving rights of use and access. This is mainly internal to an institution.
3. *Access and exploitation*: once the data is cleaned and contextualised it needs to be made accessible. This phase includes all the factors associated with use and exploitation, such as rules and regulations on its use, advertising digital services and exporting data to global aggregators such as GBIF.

Key challenges across these three stages are:

- sustainability and collaboration;
- culture;
- the right staff.

Sustainability and collaboration

Large specialised organisations are perhaps more likely to use best practice in digitisation than smaller and multidisciplinary institutions. The latter might, for instance, not be aware of global facilities and organisations such as GBIF, TDWG, CETAF and others. Furthermore, they might lack the specialised IT staff, such as data engineers and architects, who have the knowledge to make data easily portable and sustainable. However, this knowledge may be available in other institutions and national or international initiatives. Careful planning, using international standards and best practice and learning from others who have been through this process will help to avoid the pitfalls that others have experienced. There have been many collaborative digitisation projects, many of which have had a degree of success. But there are still the challenges of different institutions working in different ways to gather and share data. There does seem to be a gap between people and institutions involved in national and international projects on one side, and on the other, those managing collections on a day-to-day basis discussing the same issues that others discussed ten years previously. This is partly due to poor engagement with digitisation but is also due to staff either not seeing this as part of their career paths or not having the opportunities to engage with digital projects. Collaboration offers opportunities for smaller institutions to overcome this and ensure the sustainability of their data. Perhaps it is the responsibility of larger institutions with access to specialist staff and understanding of good practice to help smaller institutions.

Culture

The problem of sustainability relates to culture. Even though nearly all collection and research staff are now involved in digitising, willingly or not, they are not necessarily part of that culture. There are many staff who are reluctant to engage. Proof of the value of digitisation, particularly to their own projects, might bring them on board. Staff selection and having staff who can act as interfaces between the technical staff and researchers is a possible mitigation. In large institutions digitisation is core business. It does not, however, receive the financial support one might expect. In some ways, small institutions have the advantage of scale and can transform discrete datasets into virtual sets using volunteers without competing for resources with other areas of museum management. Large institutions may have big budgets but they are divided across a wide range of arguably equally important functions such as exhibitions and education. There is also a cultural variation between those who want maximum (versus basic) data on each object no matter how long it takes. For example, if the policy is to 3D digitise a collection at two objects per day, it will take considerable time before a usable dataset will be completed. During the time it would take to digitise a comprehensive collection on that basis, technologies might change considerably, potentially rendering earlier work redundant and not matching up to current data standards. A simpler dataset can be completed in far less time and allows further detailed data to be added in later. Different collections ask for different business models as basic data needed to do sound research or efficient management differ between collections. This decision might also be influenced by demand. If, for example, an external research project requires 3D scans of 100 human skulls and there is funding to hire additional staff and equipment then it will surely go ahead.

The right staff

This topic leads on from culture and culture differs between institutions, big and small, multidisciplinary or with a single focus. There is a need for interfaces between subject matter experts and data technicians and architects, if at all available. These could be seen as translators who understand the subject and have a broad understanding of the digital world. The skills needed to make digital projects happen in a *sustainable* way are different to what we are used to in museums. There are a number of challenges:

- attracting data architects and engineers away from highly paid corporate posts;
- obtaining core funding rather than relying on externally funded posts with the associated lack of stability and continuity;
- a stable core of staff with the right skill sets (see Chapter 5).

One advantage of the virtual work environment is that staff, but also volunteers, can work remotely and through the latest communications technology they can be supervised, supported and remain in constant communication with the team. Over the years, commercial companies have learned about our collections needs and have invested in specialised technologies.

Nowadays, large herbarium collections of 1 million or more sheets of dried plants are imaged, barcoded and databased faster, cheaper, safer and of a better and constant quality by commercial companies than institutions can do it themselves. A company that developed and improved its working methods with Naturalis Biodiversity Center in Leiden between 2010 and 2015 is now digitising herbarium collections in multiple European countries and the USA, showing that making use of what others did before can be a very rewarding way of completing what once seemed impossibly large tasks.

What should be digitised, how do you prioritise and who should do it?

Is waiting to digitise your collection still an option? Not being the first, avoiding the law of the "handicap of a head start", might pay off. But does it really? Do you want, if finances allow, to be a pioneer and work with others to develop new techniques for collections which challenge you to get objects digitised efficiently or do you want to be last in line? Prepare your collection in such a way that it is ready to be digitised. It is many times more costly and time consuming to digitise a collection when it is not stored systematically or in any other logical sequence. The poor state of any collection might be an argument not to digitise it as there is a risk that it will be damaged even more through the digitsation process. But is the risk worth it? If in a botanical collection the specimens are unmounted, loose and more likely to be damaged by the digitisation process than those fixed to herbarium sheets, it could be argued that the risk is worth it if the collection is going to be digitally available rather than sitting in a box safe but unused. In such instances it might be better for your qualified staff to handle it once and get it digitised than if it will be used on a regular basis by researchers who might be less careful handling these precious objects. Also think about whether to combine different processes by the same person at the same time or try to organise it in such a way that it becomes an automated routine single focus process for those who do the job, even though this might not be that inspiring for your staff and volunteers. Rotating your staff in such a production line on a regular basis so that they get to do all parts of the process, may make it more inspiring though. The New York Botanic Garden shows how to deal with this: http://sweetgum.nybg.org/science/information-management/digitization/digitization-resources/.

Exercise

Consider the advantages and disadvantages of the various options with regard to efficiency and value for money for the various collections in your institution.

If collections are vast, say between 100,000 and 1 million objects, and organisations have restricted budgets or staff resources then choices need to be made as to what should be digitised first, what later and what not at all; this should be part of your strategic plan. Arguments for digitisation can vary, even within an organisation, ranging from content-related to efficiency-related reasons: for example a choice between important historical or scientific collections which are under active study and less important collections which might be stored remotely. One argument for the latter is that it will enable collection managers and researchers to consult such collections at low cost without having to travel to off-site storage facilities. Choices can be made between basic versus in-depth digitisation, from simple databasing and imaging of labels to high quality 3D or scanning photography. Or scanning of whole boxes, drawers and jars in which multiple objects are stored allowing quick assessment of what is stored where and how without checking or opening the containers physically. It will also give visitors a preview of what you have stored before actually visiting you. Apart from these internally driven priorities, your clients in the "outside world" might request a collection to be digitised (digitising on demand) or let you decide from a strategic point of view what to digitise. Once choices have been made, it does not exclude additional choices at a later stage.

The entire process from prioritisation based on research needs, collection management priorities and external requests, through to making the data available to users can be laid out in a flow chart. Figure 7.1 illustrates the schematic overview drawn up for the large-scale digitisation project that ran from 2010–2015 at the Naturalis Biodiversity Center. The key steps can be followed: prioritisation of collections, handling of the objects, (re)identification where necessary, the practical process of digitisation (how and where), storage, indexing and finally disclosure of the virtual data.

The *what*, *how* and *by whom* questions need careful consideration within the budget available. The question of who should do the digitisation and annotation, whether it is part of the job description of museum staff, done by volunteers within or a "crowd" outside an organisation, or even executed by commercial companies, depends on factors such as size of the collection, budget available, techniques to be used and how quickly the data are needed by the users. Expert users of your digital data might be able and willing to help you by adding additional data such as geo-references to your objects as they work on them for their own study and publications. Consider what options you have and whether your institution would benefit from either a slow start involving your staff right from the beginning, making errors but learning from them as a natural process or, whether you might want to jump ahead and do what others have been doing as long as you do not forget your strategic considerations. But remember, relevance and the justification for maintaining a collection is supported by use. Today, most uses include some virtual aspect, so choosing to *not* digitise at all may be choosing to become irrelevant.

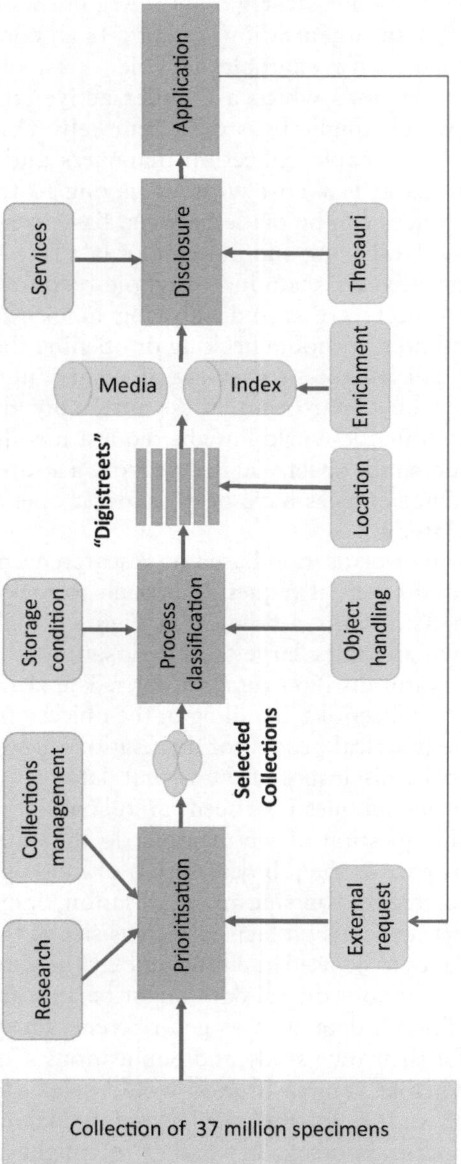

Figure 7.1 Schematic overview of the Naturalis approach to digitisation process organisation. "Digistreets" are a production line for preparing and capturing images and data

Source: Naturalis Biodiversity Center.

Exercise

- Which of the collections you manage is your preference to be digitised?
- What criteria would you apply to select this collection over other collections?
- Which of your stakeholders need to be involved?
- Who will digitise your collection and how are you going to organise it?
- Make a decision tree to document your decisions and possible alternatives.

Which standards, software and hardware should you use?

Digitised collections, whether large or small, need to be created according to common standards and be interconnectable. In the world of natural science collections, the Taxonomic Databases Working Group (TDWG/Biodiversity Information Standards, www.tdwg.org, and Integrated Digitized Biocollections, iDigBio, www.idigbio.org) are platforms for this. The TDWG community's priority is the development of standards for the exchange of biological/biodiversity data. It is a not-for-profit scientific and educational association that is affiliated with the International Union of Biological Sciences, formed to establish international collaboration among biological database projects. Their mission is to develop, adopt and promote standards and guidelines for the recording and exchange of data about organisms and promote the use of standards. iDigBio's mission is, among others, to develop an infrastructure by overseeing implementation of standards and best practices for digitisation.

As software and hardware are continuously being improved, changing rapidly over time, we are not going to advise on which to choose. Let yourself be advised by institutions who have gone through that stage and are still working on or have recently completed the digitisation of such collections; you can do this by direct contact or through appropriate networks. For reasons of efficiency and uniformity, it is wise "to follow the leader" and look at initiatives taken by CETAF, SPNHC, TDWG and others and start from there and make the necessary improvements if needed or advised based upon their experience. The same is true for data storage. There should be no competition between institutions as sharing our virtual collections is a common goal as shown by the Distributed System of Scientific Collections (DISSCo) initiative for instance.

Common principles on digital data

There are global initiatives that provide guidelines or common principles on digital data. One such initiative within the EU-funded SYNTHESYS project aimed, among others, to develop and adopt common principles for management of digital data produced by external users to balance their costs and benefits. This document, published in December 2016, is available on the SYNTHESYS website at http://synthesys3.biowikifarm.net/syn3/NA2/objective1/task1/cmp. It provides help in maintaining the highest standards for the management of digital collection data and creates a framework for the management and conservation of digital data produced by external users of natural science collections hosted by European institutions. The Wiki-Handbook *Recommendations: Management Policy on Digital Collections (MPDC)* gives advice to institutions on how to set up institutional management policies on digital collections. It lists subjects that need or can be covered in an MPDC and provides explanations as well as definitions on these. Institutions can choose the elements applicable to their digital collections in compliance with institutional as well as national regulations. It is a freely available online Wiki-Handbook, can be updated, added and amended by its users and so will continue to be up to date.

If you decide to add 3D images to your digitised data, the *Handbook of Best Practice for 3D Imaging for Natural History Collections* might help you: http://biowikifarm.net/v-mfn/3d-handbook/3d_Imaging_Handbook:About. The scope of this handbook is to present different digitisation techniques currently in use in European institutions. It provides guidelines by showing which techniques can be used for a certain collection or object. In addition to the general information on particular techniques, it provides a thorough workflow, an equipment list and a test case per institution on how to create a good digital replica of an object.

Open access

In 2019, nearly 100 natural history institutions and related organisations committed to promoting open access have signed the *Bouchout Declaration for Open Biodiversity Knowledge Management* (June 2014) (http://bouchoutdeclaration.org/declaration). The Declaration seeks to unlock the legacy of biodiversity knowledge, make digital data about biodiversity openly available and put it to work for the benefit of all. It calls for legal and technical reforms to grant all users the right and ability to freely access, copy, distribute and re-purpose published biodiversity content, accompanied by the respectful attribution of the content source as demanded by scholarly tradition. However, while open access is the default position for many museums, we still have to honour the agreements about access that were made when we acquired the object which underscores the importance of digital access to legal records so staff can verify terms of use. See also: "The FAIR Guiding Principles for Scientific Data Management and Stewardship" (Wilkinson et al., 2016).

How much does it cost?

The costs for digitisation will differ significantly depending on many kinds of variables, such as:

- simple transcription of basic data versus capturing all data including remarks and additional notes from labels;
- variable handling time, for example, from easy to handle herbarium sheets to labour intensive wet collections in jars;
- time spent on transfer of objects from storage space to digitisation lab – the closer the cheaper (and safer);
- the legibility of information on labels, e.g. hard-to-decipher 19th century handwritten versus easily-readable 20th century printed labels;
- the decision on whether or not to image objects and if so whether to choose 2D, 3D or microscopic imaging;
- the duration of the experience and efficiency gained during a project and the number of digitisers involved, making "conveyor belt-like" work possible or not.

As an example, Table 7.1 shows the average costs in euros (including overheads) calculated for digitising different types of specimens in the Naturalis Biodiversity Center in 2014.

Careful planning can save you a lot of money and will enable you to digitise much more within your budget.

Exercise

Make a cost estimate for the collection you selected in the previous exercise.

Table 7.1 Costs for digitisation: average cost in euros (including overhead) per object at Naturalis Biodiversity Center in 2014

Collection	*N Realised*	*Euro/object*
Wood samples	125,000	1.27
Herbarium sheets (Leiden)	2,600,000	1.29
Molluscs	640,000	1.37
Herbarium sheets (Wageningen)	830,000	1.47
Entomology	530,000	1.51
Microscopic slides	510,000	1.57
2D objects (books, journals, etc.)	450,000	1.87
Geological specimens	50,000	1.90
Vertebrates (dry)	170,000	2.37
Specimens preserved in alcohol	60,000	4.65

Crowd-sourcing as an example

In 2013, Naturalis Biodiversity Center in the Netherlands experimented with online digitisation of collection data through crowd-sourcing. The aim of the project was to have labels of 100,000 glass slides transcribed and validated by a community of online volunteers within a time span of six months, at a cost comparable to that of the internal process of digitisation of glass slides. The project was carried out on an existing online platform for transcription of written heritage: https://velehanden.nl.

Each glass slide was transcribed by two different volunteers to check for and avoid errors and validated by a third volunteer, amounting to 300,000 treatments in total. On average, the labels of 711 slides were transcribed per day with a peak of 1,913 slides. The quality of the work was very high and did not differ from in-house digitisation. Nevertheless, the project duration was exceeded by three months, mainly due to an overly optimistic estimate made prior to the project.

A total of 497 volunteers signed up during the project with a broad demographic spread and the majority aged over 55. The really active group consisted of only 203 volunteers, while approximately 60% of the whole volunteer force transcribed less than 10 slides. Of this active group, 26 individuals were responsible for 90% of the work and the two most active members together accounted for over 68,000 transcripts. This limited number of active volunteers may reflect cultural differences and perhaps different results would have been observed in countries where volunteering was more embedded in society.

The total cost of the project was 33% higher than the originally estimated 60,000 euros, mainly because it lasted for nine instead of six months. The cost per slide turned out to be 0.79 euro which is 9 cents higher than in-house transcription.

Given the limited scope of this pilot and the somewhat specialised target group, the recommendation was to investigate how crowd-sourcing activities might attract a more diverse audience. A good example is provided by the "herbonautes project" of the Muséum National d'Histoire Naturelle (MNHN) in Paris. Instead of transcribing objects at random without a direct relation to one another, as in Naturalis, the MNHN stimulated volunteers to be responsible for a specific but limited part of a collection, giving volunteers a sense of ownership and responsibility for "their" section (see http://lesherbonautes.mnhn.fr and www.researchgate.net/publication/295688898). The "herbonautes project" has been implemented at other herbaria, e.g. the Botanischer Garten und Botanisches Museum Berlin (BGBM) (see www.herbonauten.de/).

What digital initiatives exist around the globe?

The rapidly increasing number of national and international initiatives presenting and linking biodiversity data on the web makes an extensive overview of such initiatives out of date as soon as it is published. The reader is advised to search the web to see what is available and what is new, from early initiatives, such as the Global Biodiversity Information Facility (GBIF), VertNet and Europeana, to the more recent Innovations Consolidation for Large Scale Digitisation of Natural Heritage (ICEDIG) and Distributed System of Scientific Collections (DISSCo). The aim and mission of these initiatives as extracted from their websites are given below as an example.

Examples of large-scale digitising initiatives

GBIF (www.gbif.org): "The Global Biodiversity Information Facility is an international network and research infrastructure funded by the world's governments aimed at providing anyone, anywhere, open access to data about all types of life on Earth. GBIF provides data-holding institutions around the world with common standards and open-source tools that enable them to share information about where and when species have been recorded. This forms the basis of GBIF's index of hundreds of millions of species occurrence records".

VertNet (www.vertnet.org): a "collaborative community effort to bring together the expertise of biodiversity scientists and information experts making biodiversity data free and available on the web. VertNet is a tool designed to help people discover, capture, and publish biodiversity data. It is also the core of a collaboration between hundreds of biocollections that contribute biodiversity data and work together to improve it. VertNet is an engine for training current and future professionals to use and build upon best practices in data quality, curation, research, and data publishing".

Europeana (https://pro.europeana.eu): an online collection of millions of digitised images containing material, including natural history, from European museums, libraries, archives and multimedia collections.

ICEDIG (https://icedig.eu): "The Innovation and Consolidation for Large Scale Digitisation of Natural Heritage is an EU-funded project that aims at supporting the implementation phase of the new Research Infrastructure DiSSCo (Distributed System of Scientific Collections) by designing and addressing the technical, financial, policy and governance aspects necessary to operate such a large distributed initiative for natural sciences collections across Europe".

DISSCo (http://dissco.eu/): "The Distributed System of Scientific Collections is a pan-European Research Infrastructure initiative with a vision to position European natural science collections at the centre of data-driven scientific excellence and innovation in environmental research, climate change, food security, health and the bioeconomy. 115 organisations across 21 countries have already committed to DiSSCo. Its mission is to mobilise, unify and deliver bio- and geodiversity information at the scale, form and precision required by scientific communities: transforming a fragmented landscape into a coherent and responsive research infrastructure".

What are the risks?

Digitising collections and creating a virtual "new" collection is not without challenges. There are many factors which will have to be taken into account in the planning process. First of all, virtual collections do not replace physical collections. In fact, it adds a new collection to your collections: the virtual collection.

Some, both within and outside institutions, still believe that once a collection is digitised, the physical collection can be discarded (see Chapter 2). This might be the case with collections such as newspapers or journals which are not unique, but not with natural science collections consisting of unique individuals and samples or with man-made objects such as artworks. Virtual collections do not allow for analysis of, among others, DNA, isotopes, chemicals, minerals, materials used or anything else we might be able to discover and examine in the future.

Virtual and physical collections are different types of collections which need their own strategy and management. Furthermore, virtual collections require specialised staff, IT support, software, hardware and storage facilities and hence imply significant, structural costs. If not well planned, pitfalls lie within each of these components. They will differ between organisations of different sizes and collections based on budget, capacity and opportunities. This does not mean that only institutions with large budgets or status can afford to digitise. Smaller museums can draw on volunteers and the crowd to capture data for use within their institution and local user community, but can also make their data available to the wider community by contributing data to aggregators such as GBIF. To all organisations large or small there is a further, less obvious risk. In this era of digital transformation, you may be risking your collection becoming invisible, unused and irrelevant by not ensuring digital access. This could result in less funding or even closure of collections and institutions.

One general issue which is yet to be resolved is the continuing addition of annotations, such as new names to specimens by scientists and the "crowd" to databases and virtual collections. Visitors to the collections and borrowers of physical specimens are mainly experts in their group of plants and animals and add valuable data on current names, misidentifications and type status on attached paper slips for example. The same specimen may or not have information updated on either or both its physical or virtual representation. Virtual visitors may or may not return changes and there is often no mechanism for them to do so. As more users consult virtual rather than physical objects the problem will get worse. The risk here is that the physical collections will become out of date and of less value and accessibility. The information users get from the virtual object will therefore differ from what one gets from the physical object which lacks these annotations. The problem is magnified if the digital records available through portals are not kept current with the institution's digital records. Solutions to this can only be reached at an international level, by multiple parties involved.

There is always a risk of inaccurate data being added. If you plan to digitise your collection through crowd-sourcing, how do you mitigate against this and ensure accuracy? How do you avoid the introduction of errors and inconsistencies by your crowd? There are several ways to validate or cross-check these data, as illustrated in the section "Crowd-sourcing as an example" above.

Other risks include:

- Damage to and loss of objects during the digitisation process.
- A shift of attention to virtual objects resulting in less routine inspection of physical collections through access and less detection of insect pests for example. These risks are balanced by possible reduction in handling damage.
- Dilemma of division of distributing limited resources between physical and digital collections resulting in suboptimal care for both.
- The link between the physical object, the voucher and digital object becomes diluted or lost. The digital object will "live its own life" no longer linked to the original object resulting in separation of information between the two.
- Unaffordable costs of updating and maintaining data leading to "orphaned" and out of date virtual collections.
- Data being released without authorization.
- Obsolete media and digital records being held on unsustainable platforms.
- Highly unique data structures that prevent or slow down aggregation making digitisation efforts futile.
- Inability to meet users' needs for data because of inadequate planning.

Exercise

What are the pitfalls regarding digitisation for the collection you selected and the choices you made in the previous exercise?

What is the future for digitisation?

Much of what was considered state of the art in digitisation only a few years ago, has improved, been made more efficient and has become cheaper. Waiting to digitise is no longer an option. Do not wait for it to become even cheaper as you will miss opportunities, partnerships in international projects, visibility, etc. Digitisation is accelerating and becoming more cooperative. It has become one of the top priorities of museums and herbaria and not only in natural sciences. Data quality will continue to be improved and additional links will be created between objects and whatever is considered essential to these objects. Links can be made between the virtual objects and publications and websites in which these objects have been illustrated and described going back 200 years or more. Links can be made between virtual objects and additional information from whatever sources about the object, collector, artist or geographic origin, the latter for instance via websites that provide satellite views. Virtual reality technologies have created virtual exhibitions, based on high-quality 3D models (Bruno et al., 2010), while artificial intelligence and machine learning are "very powerful tools and are more accessible than ever before. In the hands of museums, these technologies will inevitably lead to interesting discoveries, rich data, and new paths into your collection" (Ciecko, 2017). Options are multiple and seemingly endless, illustrating the ever growing discrepancy between enriched virtual objects versus much more static physical objects.

Summary

The increasing demand for collections data in web-accessible digital form is creating new virtual collections which, like the physical collection, require careful management and curation to ensure currency and long-term value. Data can be of varying quality and one major challenge is to ensure that your data are of such quality to be usable by the community.

Virtual collections are not a replacement but an addition to the physical collection. They are collections in their own right and perhaps need to be treated as such. Virtual collections are a great benefit to collection management tasks and improve access to collections, raising it to a new level offering new opportunities to use and link collection information. Compared to physical collections, to which access is controlled, online virtual objects can be annotated by a crowd of anonymous users linked only to the virtual, but not the physical, object. There are associated risks; for example discrepancy in information linked to the same object causing possible misinterpretations or worse, errors in scientific communication. However, the additional data linked to a virtual object greatly increases its research value.

Planning digitisation projects to guarantee optimal access to your data requires complex decision making on methods, standards and legal issues. Resourcing is a key issue and options such as crowd-sourcing, dedicated digitising teams and new technologies are reviewed. Fortunately there are many examples of digitising projects available from which you can learn.

The key strategic questions, which can be different for each institution, are: why should digitisation be carried out; to what level should it be done and on which collections?

Start simple, step up production when you can, add and improve later. Visibility of your collection can come first by broad and shallow data gathering. You can prioritise which collections need to be digitised in more depth later on the basis of end-users (of which you as holder of the collection are one) – demonstrating what is really of importance. This can be from a scientific, public, educational and managerial point of view. If you need to add images, add images. If you need additional data, add additional data. And if certain collections prove to be important, you might be able to generate additional funds more easily and from unexpected sources, even crowd-funding.

References

Bruno, F., Bruno, S., De Sensi, G., Luchi, M.-L., Mancuso, S., and Muzzupappa, M., 2010, From 3D Reconstruction to Virtual Reality: A Complete Methodology for Digital Archaeological Exhibition. *Journal of Cultural Heritage*, 11: 42–49.

Ciecko, B., 2017, Examining the Impact of Artificial Intelligence in Museums. *MW17: Museums and the Web 2017*. https://mw17.mwconf.org/paper/exploring-artificial-intelligence-in-museums/ [Accessed 15/0819]

Wilkinson, M. D. et al., 2016, The FAIR Guiding Principles for Scientific Data Management and Sewardship. *Scientific Data*, 3: 160018. (doi: 10.1038/sdata.2016.18)

Further reading

Ang, Y. et al., 2013, A Plea for Digital Reference Collections and Other Science-Based Digitization Initiatives in Taxonomy: Sepsidnet As Exemplar. *Systematic Entomology*, 38, 637–644 (*and references therein*).

Blagoderoc, V., and Smith, V.S. (eds.), 2012, *ZooKeys 209: No Specimen Left Behind: Mass Digitization of Natural History Collections*. Sofia-Moscow: Pensoft.

Enghoff, H., 2017, Does Digitization of Natural History Collections Reduce the Need for Physical Access and Physical Loans? Report on Subtask 3.2.1 Under Task 3.2: "Facilitating Access beyond SYNTHESYS3".

Heerlien, M., Van Leusen, J., Schnorr, S., Jong-Kole, S. de, Raes, N., and van Hulsen, K., 2015, The Natural History Production Line: An Industrial Approach to the Digitization of Scientific Collections. *Journal on Computing and Cultural Heritage*, 8 (1): 3. (doi: http://dx.doi.org/10.1145/2644822)

Leusen, J. van and Heerlien, M., 2012, Digitaliseren van de entomologische collecties van Naturalis Biodiversity Center. *Entomologische Berichten*, 72 (5): 259–262. www.repository.naturalis.nl/document/453900/ [Accessed 15/0819]

van den Oever, J.P., and Gofferjé, M., 2012, "From Pilot to Production": Large Scale Digitisation Project at Naturalis Biodiversity Center. *ZooKeys*, 209: 87–92.

8 Access to collections

Introduction

Some of the key tasks of organisations holding natural science collections are to help in answering scientific and societal questions and to provide the basis for public engagement activities. Access is the key to keeping the collections relevant. Access is not one-way. Use adds value to the objects by connecting them with information, placing them in another, new context that builds the basis for new questions and uses. It should be in the core interest of any organisation holding collections to encourage and facilitate access and open the collection up for a broad range of uses and user groups. As museums and research institutes, it is part of our mission to support use and give access to our collections.

On the other hand, we are responsible and accountable for the protection and preservation of the objects we are managing. The goal for all collection professionals, whether in the arts and humanities or natural sciences, should be to enable access while minimising risks. The framework in which access is allowed is set by (inter)national rules and regulations, institutional priorities, laws and policies and by conditions such as staff resources and structure, and the accessibility of the collections. General principles help to deal with different types of uses and to decide on whether and how collections might be used and by whom. As managers responsible for access we are faced with the challenges of barriers, both new and longstanding, and new trends.

Questions

- What is access to collections?
- What are the restrictions on access and who is responsible for allowing access?
- What are the principles for physical access to collections?
- What are the principles for digital access to collections?
- What are the benefits of access and how can we measure them?
- What does access cost and how can those costs be reduced?
- What are the "right" decisions to make on access?

- What are the major barriers to access and how can you overcome them?
- What will access look like in the future and how can we be prepared?

What is access to collections?

Authentic physical objects are the core of our activities as research institutes and public museums, be they scientific studies, student courses or exhibits. Access to collections comprises all activities from research and education to public engagement. Access to objects and related information is provided to enable their manifold use. Access can be given directly to the physical objects by individual visits or loans (physical access) and to (meta-) data and information linked to them, e.g. through requests or digital access to online databases. In recent years, there has been a shift from physical to digital access. Still, enabling physical access to the objects remains one of the central services of organisations holding natural science collections. Very specifically for these kinds of collections, there is a strong focus on research.

As an example, the Museum für Naturkunde in Berlin, Germany, holds about 30 million zoological, paleontological and mineralogical objects. On average they are visited by 600 external visitors, mainly researchers, each year. Altogether, these visitors spend almost 2,000 days per annum in the collections. Almost 40,000 objects, mainly insects, are sent on loan in more than 600 loan procedures. 80% of the successful loan requests are for research purposes, 20% for exhibitions. 90% of the objects are sent to researchers for scientific investigations, only 10% (about 4,000 objects) are part of loans for exhibitions. Whereas the annual variation in scientific loans is rather low, the number of loans and objects for exhibits changes considerably between years depending on ongoing exhibition projects.

Access versus use – a brief explanation

- Physical access: access to the physical objects through visits and loans.
- Digital access: access to object information via databases, images, scans, etc.
- Uses: research, display, education, art, commercial.
- Users: internal/external researchers, amateurs, general public, companies, agencies.

Use of natural science collections, particularly in larger organisations, focusses mainly on research and includes comparisons of morphological characteristics, sampling, databasing and imaging. Exhibitions of objects in public galleries or open store depots enable the public to access the physical collections. The public display of objects is probably the most restricted and controlled way of physical access as collections security and health and safety issues are primary. Physical access puts the objects at risk through direct agents such as poor handling, sampling, loss or theft, or indirect agents such as changes in storage conditions when specimens are removed from storage. Users themselves can also be at risk as most biological collections have been treated with biocides such as mercury, arsenic or naphthalene in order to protect them from pests. It is important that visitors and borrowers are aware of these risks and are protected by the right safety precautions. Legal aspects and regulations, such as access and benefit sharing (ABS), health and safety and transport of dangerous goods, are setting a framework for deciding whether access can be provided or not (see Chapter 4).

What are the restrictions on access and who is responsible for allowing access?

As we discussed in Chapter 2, most natural science collections are research infrastructures, and as such they work by the general principle of ensuring and facilitating access to a wide variety of uses and users. This not only includes colleagues and scientists from your own and other organisations but also amateurs, artists, agencies and commercial companies. Collections are there to be used unless specific restrictions apply. This presumption of a manifold barrier-free access (easy, timely, convenient and customised) is in contrast to other disciplines like arts or humanities. This is partly linked to the fact that the main value of natural science objects evolves from their use, information and the knowledge that we build around them. However, we should remain aware of the monetary value of objects which might attract the wrong kind of "users" (see Chapter 6).

Decisions on access should aim at maximising use while keeping risks to an acceptable limit. Depending on objects and topics involved, access and use might be restricted, withdrawn or not allowed in the first place. Reasons for this include failure to follow ethical codes for museums or other national and international regulations or high risk of damage or theft to fragile and valuable objects.

Valid reasons to restrict, withdraw or deny access to your collections

There are some valid reasons to restrict, withdraw or deny access to your collections. These include:

- Use for commercial purposes without permission;
- Failure to follow ethical codes for museums and relevant collections, e.g. ICOM Code of Ethics for Museums;
- Failure to comply with national and international agreements and regulations, e.g. on access and benefit sharing, CITES, transport of dangerous goods etc.;

- Use of objects in unacceptable contexts and under damaging conditions, e.g. natural science objects in art installations, decorations, etc.
- Exceptionally high risk of damage or theft of fragile, light sensitive, rare and/or valuable objects;
- Exceptionally high risk through inappropriate climate conditions (light, humidity, temperature) and lack of security measures during use;
- Sensitive data, e.g. distribution data of rare and valuable species (targeting by thieves!);
- Disproportional/exceptional cost-benefit ratio, limited resources;
- Access interferes with programmed curatorial or research work.

Decisions on access to collections are not always easy but are influenced by several, sometimes contradictory factors, e.g. conservation requirements, scientific background and legal aspects. Expertise in these areas is often held by different staff members such as researchers, conservators and collections managers. Decision-making processes should make sure that the staff responsible for the collections are involved and empowered through management structures and job descriptions to take these decisions. Processes and responsibilities vary between, and sometimes even within, organisations. In larger organisations, heads of collections, curators or senior researchers are responsible for the scientific collections, but often not for collections for educational programmes and exhibitions and "research collections" arising from specific research projects. They often fall under the responsibility of the respective departments or individual researchers. On the other hand, in small institutions all these functions might be managed by a single person. Decision-making processes can work well in all management structures as long as the processes fit to the management structures of the organisation and are well-supported by necessary competencies (see Chapter 5).

Do you think decision making on access varies between institutions with different management and staff structures? Take, for example, Museum A in which research and collection management is fully integrated and where decisions on access are taken by a (research) curator, and Museum B with a separate department of collections where decisions are taken by a head of collections (see Chapter 5). Would criteria to allow or restrict access be the same in these two institutions? For example, could the curator prevent access to a collection that he or she is working on? Would the head of collections decide similarly? Could the collections manager prevent access that might risk the integrity of a specimen no matter what discoveries might be made and would the curator ever use this as an argument?

Exercise

Thinking about a collection with which you are familiar, what might be valid reasons to restrict, withdraw or deny access?

Regardless of management structures, institutional policies and procedures need to be transparent and clear about who can have access, for what purposes, to which objects and data and under which conditions, which restrictions, and whose responsibility. Transparency and proper documentation are the key factors for efficiency of internal processes as well as clarity for users. Criteria for access and decision-making processes should therefore be on every institutional website to enable visitors to decide in advance whether their requests are likely to be approved or not.

Individual organisations, as well as networks, have built up a lot of expertise and best practice on access. Before developing policies and procedures, you should feel encouraged to use expertise and knowledge held by community networks such as CETAF (www.cetaf.org) and SPNHC (www.spnhc.org). An excellent example is the Code of Conduct for access and benefit sharing developed by CETAF (https://cetaf.org/services/natural-science-collections-and-access-and-benefit-sharing). Apart from background information, it contains principles and practices for implementation and useful templates.

What are the principles for physical access to collections?

Policies and procedures are often guided by, or even determined by, national and regional laws and regulations and legal frameworks given by the funding body. There are many regulations that relate specifically to biological collections. Furthermore, policies and procedures must fit into the institutional management structure and its strategic vision, mission and priorities. There is no "one-size-fits-all" solution. However, general principles based on international rules and regulations and good practice can help in developing policies and adapting them to specific situations. We have listed a number of such principles below. Most are applicable to all aspects of physical access such as visits, loans and destructive sampling and are independent from organisational structures.

General principles of physical access

Policies and procedures concerning access should aim at stimulating use of the collection and add information to objects while preventing damage or loss of these objects.

General terms, conditions and restrictions for access should be available to all users, staff, visitors and borrowers beforehand, preferably on the institution's website. Simplified procedures might apply for internal users such as education and exhibition departments. Before touching objects, users should be informed on the handling of these objects with regard to health and safety, conservation issues and specific requirements. Terms and conditions should also be clear about restrictions such as:

- Forwarding to third parties is only allowed after approval by the owner.
- No changes to original labels are allowed. Annotations should be provided electronically, by determination labels or any other form of hard copy.
- Physical changes to objects such as by sampling, dissection, or repair of pedestal/armature or sheets are only allowed after approval by the owner.

Unless specific restrictions apply, all objects should be available to researchers in any discipline, for exhibitions, art installations and educational activities in-house or at partner institutions, as well as for development of new technologies and methods.

For most public organisations, collections are not primarily meant to be used for commercial purposes. However, as part of collaboration and innovation projects collections might be used for commercial reasons and benefit from additional resources and investment derived from earnings. In these cases, specific agreements (contracts) are necessary to avoid clashes with the mandate of the organisation and to ensure mutual benefit.

Organisations reserve the right to refuse access, cancel or deny visits and loans. Arguments should be transparent. They may include excessive costs, unannounced visits, emergency situations, unacceptable risk to fragile or sensitive items (e.g. types, objects of high historical significance or hazardous materials), legal restrictions, and failure to comply with previous loan agreements or others.

As a security measure, a careful documentation of all users and uses and related processes is essential. Documenting and archiving personal data must be in compliance with the General Data Protection Regulation (GDPR; https://eugdpr.org/). Documentation of access to collections comprises the following areas:

- correspondence, including initial requests for visits and loans stating name, affiliation, scope and aim of the work and a specific description of the material needed;
- procedures of approval and arrangement of visits or loans on the basis of prior agreed terms and conditions, e.g. agreements, contracts, agreed terms of use and copyright agreements for photos;
- a visitors' book or form including name, address, affiliation, date and purpose of the visit and material handled;
- a loan contract stating name and address of the borrower, identification of the specimens (if necessary with photos), expiry date and signatures by both parties.

The Collection Policy Board, a working group on decision taking at management level under the umbrella of CETAF developed and agreed upon common principles for research loans and visits to the collection. These documents are included here as Appendices 8.1 and 8.2.

Principles of visits to the collection

Visits to the collection need to have a clear purpose and be prearranged as laid out in the policies and procedures and on the website. Requests for visits should make clear why access is needed, what the aim and scope of the work is, what specifically will be done with the material, what kind of resources (staff, labs, etc.) and time in the collection are needed. This information builds the basis for the decision taking and arrangements.

A series of past thefts has shown that a careful check of the personal identity (name, address), and wherever possible passport or other recognised ID cards, of visitors when arriving can be essential. Visitors should always be assisted with expertise and technical advice during their stay. Bags, coats and other items that might be used to hide and thus steal objects should not be allowed in collection storage facilities or laboratories. At the end of the visit, a careful check by staff should make sure that all objects are still there and in good order.

Example: The feather theft story

In 2017 a number of articles appeared in the media (for example www.herald.ng/swiss-man-trial-stealing-feathers-natural-history-museums and www.thelocal.ch/20170703/swiss-birdwatcher-charged-with-stealing-feathers-from-museums) regarding the court appearance in Basel, Switzerland of a bird enthusiast whose ambition was to own a feather from every species of bird of prey in the world. The man had admitted to stealing 458 feathers and wings from museum ornithological collections in Switzerland, Germany and Austria, resulting in significant damage to often rare and scientifically valuable bird specimens amounting to almost 5.4 million euros. Museum staff had granted the thief access believing him to be a bona fide ornithologist who needed to study specimens to complete a scientific paper. In 2012, after several years of undetected pilfering, staff at the Museum für Naturkunde, Berlin spotted damage to a bird specimen and through examining their visitors' book made the connection to the visit of the thief to the collections. A survey of specimens instigated by the bird collections community revealed that the man had used this same method of access in several museums to amass his hoard of stolen material.

Principles of loans

Policies and procedures regarding loans should ensure transparency of the process, especially the traceability of collection items, clarity of responsibilities, acceptable storage and handling conditions at the borrowing institute, and agreements on collection use, including legal requirements such as regulations on access and benefit sharing (ABS).

Sending objects on loan involves a risk of loss and damage. To avoid loss of entire collections, only a proportion of the institution's holdings of a collection unit should be sent on loan at any one time. Furthermore, the duration of a loan should be clearly restricted in time. Normally this varies between 6 and 12 months. Extensions and exceptions can be agreed upon between parties.

Material should only be sent to organisations that hold collections and hence have suitable storage facilities and staff responsible for their management. Loans to individuals at private addresses is common in entomology with a huge community of amateurs. Specimens face numerous dangers such as the absence of pest control and the chance of fire, theft, loss during house moves or even death of the borrower. Loans to private addresses should only be allowed, if at all, under exceptional circumstances and specific terms.

Between partner institutions, normally no charges for packing material, postage and staff time will apply. Exceptions can be made for repairs to, or replacement of damaged specimens, for specialist packing, treatment or any unusual or exceptional means of carrying, or in the case of extensive requests or additional (external) services.

Specific comprehensive requirements might also apply for loans for exhibitions, especially regarding insurance, transport, safety during exhibition and documentation of the loan. It is important to inform partners of these costs beforehand.

Principles of destructive sampling

Some research questions can only be answered through analyses that cause irreversible changes and damage to collection items. Known as "destructive sampling", it includes any procedure that causes a permanent change to a specimen, such as sampling of tissue for molecular studies, dissection of genitalia of insects, flowers from herbarium plants, slides from geological samples, or tooth enamel or bones for isotope studies. It also includes some reversible procedures, such as disarticulation of mounted specimens or separation of elements that have been glued together. Requests for destructive sampling need careful consideration. Apart from the information mentioned above, applications should include statements on the sampling methods and reasons why a certain protocol is chosen and evidence that this protocol will be successful, e.g. by using the online tool "thermal-age.eu". Thermal-age.eu helps to predict the likely survival of DNA and other molecules in archaeological and fossil samples.

A coordinating mechanism for destructive sampling by a knowledgeable, dedicated curator with a high understanding of the balance of risk and benefit or, alternatively, a panel will enable a transparent decision to be made, especially when it comes to:

- feasibility of success for the destructive sampling request;
- minimisation of damage or over-use of samples;
- prioritisation of loans in the case of multiple identical requests and communication with parties involved in these requests;
- review of exceptional circumstances, e.g. a request which will destroy the entire object.

Some organisations might ask applicants to use their own on-site facilities to reduce further risks to the specimens and ensure that certain standards and protocols will be met. In that case, they will take care of unused fragments or other products of the sampling process. In general, any fragment remains the property of the owner and must be returned to the institution of origin. If destructive sampling leads to other products (such as slides or geological objects), and the borrower is allowed to keep them, the institute of origin has the right to ask for duplicates.

What do we expect from users?

While the process of access to collections is clearly defined and users are expected to subscribe to and follow terms and conditions, feedback from them is not. Many users inform the organisation of origin about the results and findings of their activities. However, this is organised in many different ways and there is no common or formal way yet to force visitors and borrowers to report back. In order to increase visibility and relevance of collections, it should be part of any access policy to encourage them to do so and provide information (annotations, data, images or other media) and publications based on the specimens used. In these publications, the institution of origin must at least be acknowledged by using the official name and reference to the specimens should be given through their unique identifiers such as registration numbers, collectors numbers, etc. wherever possible.

Within the European project SYNTHESYS, the Collection Policy Board developed common principles on digital data produced by external users in order to value the contribution of external users to the development and disclosure of the collections while ensuring the implementation of data standards, long-term storage and legal requirements (see Appendix 8.3).

What are the principles for digital access to collections?

Digital access to collections has clear advantages in many respects. It reduces the risk of objects from handling and transportation. It is time-efficient and sustainable since neither the object nor the user need to travel. By giving access to objects and information through online databases and portals multiple users can have access simultaneously, for different purposes and with different questions. Internet portals like GBIF (www.gbif.org), Europeana (www.europeana.eu) and Digital Collections Australia (www.digitalcollections.org.au) are linking data, images and information from many collections and other sources that are physically distributed. By bringing together and presenting information from different locations, they make the study of our collections directly and indirectly more efficient, e.g. by helping to prepare visits and loans.

While from the long-term perspective digital access has clear advantages, the process of digitisation puts objects at risk of damage. Individual requests for digital information should always undergo risk and cost/benefit analyses. Digitisation of objects on a larger scale (mass digitisation of comprehensive collections and millions of objects) is one of the biggest challenges due to the sheer number and vulnerability of the objects undergoing (semi-) industrial mass digitisation processes designed for standardised objects and external staff not experienced in handling fragile objects (see Chapter 7).

Public organisations generally agree on an open access policy concerning standard data (collecting data, identification) and full body images (see Bouchout Declaration for Open Biodiversity Knowledge Management, www.bouchoutdeclaration.org). Data and imaging policies are setting the framework for future use of the digital information (see Chapter 7).

Common principles for digital access and open data are the basis for common standards in sharing data and information and the development of joint e-infrastructures. A useful set of such principles is provided by the FAIR system, which aims to make data Findable, Accessible, Interoperable, and Re-usable (www.force11.org/group/fairgroup/fairprinciples). However, organisations reserve the right to restrict access to collections data and information. This is important in the case of nature conservation issues (distribution data and localities of rare or endangered species), privacy laws, intellectual property rights and other legal embargos. Databases that allow differentiated access to data following different security levels can help to keep control of sensitive data and information. That way, the public can be provided with a more general overview, whereas staff members responsible for the collections are allowed full access to all data. Processes of and infrastructure for digital access are very dynamic and developing rapidly. We would highly recommend you to get in touch with experts in your community. For natural science collections, examples include TDWG (www.tdwg.org), iDigBio (www.idigbio.org), and CETAF (www.cetaf.org). They will help you to design your own framework and to answer specific questions. In many cases, it will be the most efficient and sustainable solution to join an initiative rather than creating structures and rules yourself.

What do we expect from users?

Virtual collections are created on the basis of data associated with specific objects and collections, but they soon develop a life of their own. Even though users have signed up to specific user agreements when downloading data and media, an organisation cannot control the further use of them. However, in order to maintain the link between virtual and physical collections, it should at least be mandatory that users maintain and present a clear reference to physical objects and the organisation of origin wherever they use the data and media objects.

What are the benefits and how can we measure them?

The use of a collection can be seen as the beginning of a value chain. It starts with the handling of the objects. Extracting and digitising data and running analyses immediately add value to the objects. More value will then be added through publications. Information made available to a broader community will be included in a range of activities: scientific and public debates, new regulations and standards, technological innovations, exhibitions, art installations, public events and more. Information will be added to the objects over and over again, putting them into contexts, adding value and forming the starting point for another value chain. Figure 8.1 shows a simplified version of this value chain and a detailed discussion of the value of collections can be found in Chapter 2.

Taking this a step further, Figure 8.2 illustrates how over the last 100 years, fossils have been used and involved in many activities adding information to them and putting them into new contexts again and again, building a net of more and stronger ties between our cultural and natural heritage.

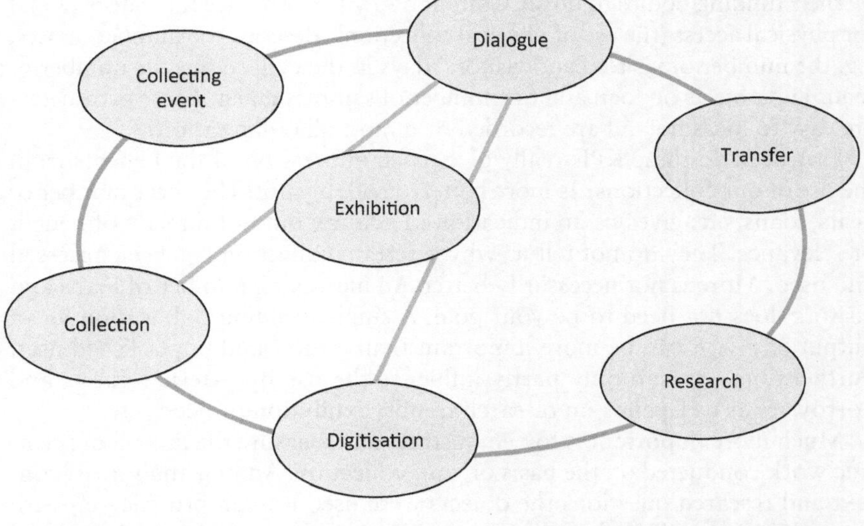

Figure 8.1 Value chain of an object from discovery to preparation, identification, research, storage, digitisation, exhibition and transfer of knowledge

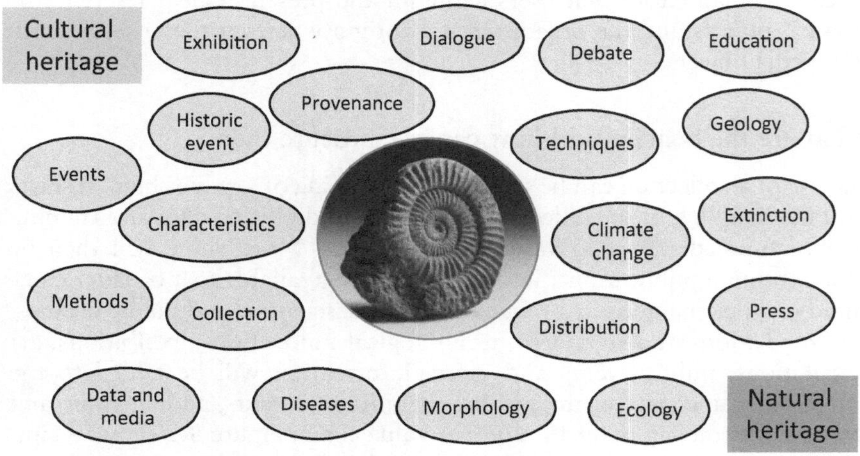

Figure 8.2 Value net for fossil specimen. A single fossil can be an inexhaustible source for research, public engagement activities, and political debats, linking disciplines and adding value
Source: Ammonit_c_Carola_Radke_MfN_Berlin

How can we measure benefits from the use of collections?

It is not only in the interest of organisations to measure, demonstrate and increase the relevance of their collections (see Chapter 2), they are also asked by their funding bodies to do so. Common Key Performance Indicators (KPIs) for physical access (the use of physical collections) describe the number of uses, e.g. the number of visitors and visitors' days in the collections, the number of loans and objects on loan and the number of samples taken. These parameters are easy to measure and are recorded by almost all organisations.

However, do these KPIs really give us an impression of the benefits from the use of our collections? Is more better? Probably not. The sheer number of visits, loans, etc. gives us an indication of activity but not directly of benefit or relevance. They do not tell us why a certain collection has been accessed and used. More is not necessarily better. An increasing number of loans and visitors does not need to be your goal. A single trending publication as an output of a visit can be more important than a thousand unpublished uses. Furthermore, we can only partly influence the use by external guests and borrowers as it depends on research trends, exhibition projects, etc.

Much more important is to demonstrate and measure the impact of scientific work conducted on the basis of your collection. Adding analyses of context and research questions the objects were used for can provide you with examples of specific value chains and improve your understanding of the strengths of your collection and potential impact and relevance. However, to enable you to do this, you must be aware of this scientific output and your institution and collection items need to be referred to correctly, preferably by using unique identifiers. In the case of more comprehensive contributions, co-authorship is an appropriate way of acknowledging your input and promoting your institution and collection.

The same is true for virtual collections. The number of database users, downloads, etc. are common KPIs. Usually "the more the better" rule applies, especially when it comes to the use of databases and information that are freely available online. This becomes even more important in the era of digital transformation (see Chapter 7). However, the sheer number of users does not say much. It would be much more interesting to get a better understanding of the users. What are data and information used for? How are they contextualized and interlinked with other information? Analyses are quickly established in an online world. However, they must be both ethical and legal, and especially in accordance with the General Data Protection Regulation (GDPR).

On the public side, the number of visitors to an exhibition or tour behind the scenes is a KPI. Often this number is directly (entrance fee, shop sales) or indirectly (proportional additional governmental funding) linked to a monetary value. Both aspects, the sheer number of visitors as well as quality and longer-term impact of their visits are similarly important. Within the limits of an organisation, the more the better is the clear aim. But how has the visit changed the life of the visitor? Did they enjoy it? What did they learn? Will they recommend your museum? Will they come back for another experience? Will they even start as a volunteer?

Other common and meaningful KPIs for access and use of your collection are the number of projects and the amount of external funding arising from and based on the use of your collections (see Chapter 3 for a broader discussion of performance indicators).

If activities and money resulting from the use of your collection can directly be spent on the improvement of the collection, this will be a powerful and direct return of investment. All users, visitors as well as borrowers, should be made aware of the importance of their work and that each of them can contribute to demonstrating the strength and importance of your collections.

What are the costs of access and how can they be reduced?

Even though most users will not be charged, access to collections is not gratis but incurs direct and indirect costs. Direct costs are directly accountable to an object. They include expenses for packing, mailing and courier services, costs for external services, etc. Indirect costs include administration, personnel, maintenance and security. In the case of access to collections, they are mainly the result of staff time spent on requests from visitors or borrowers. Costs vary widely between organisations but can easily reach several hundred euro or dollars per day, especially in the case of comprehensive loans or requests that include preparation work and sampling. Costs can be a critical factor when taking a decision on access. There is a balancing point where selecting, packing, sending and after return checking and storing the objects become more expensive for your organisation than the cost to a visitor for travel and accommodation. An invitation to visit your institution instead of sending the material on loan can reduce your direct costs considerably.

Exercise

What are the access costs in your organisation? Set up a list of cost intensive items and calculate how much a visitor's day in one of your collections would cost. Compare this to what it would cost to send the same items out on loan. What would be your decision from a cost perspective in this particular example?

Some institutions charge bench fees to visitors. Public organisations will usually not charge costs for loans or visits and if they do these costs may be reduced or waived when balanced against value added to the collection. Providing services to society, be it for researchers or the public, is part of their mandate and covered by their core budget. However, if the visiting party gains direct financial benefit from their visit, such as a painter or photographer or film producer selling its product or a commercial party making money, there is no reason not to charge bench fees. Often there are specific procedures for these kind of visitors, including a contract that clearly states the further use of data and information and costs involved. Costs might also include staff time, as these visitors are usually not trained in handling natural science objects and need full time assistance from your staff.

Private museums, on the other hand, depend on their own income and therefore might generally need to ask for bench fees.

Example: Bench fees

The Bernice P. Bishop Museum on Hawaii is a private, non-profit organisation that is required to generate revenue for costs realised in association with collection management activities. Costs may include, but are not limited to, facility bench fee, personnel time and other direct costs associated with the specific activity. Current costs include a bench fee of $200 a day and staff time at $75 an hour. Fees may be discounted or waived depending on circumstances (see www.bishopmuseum.org).

Regardless of the mandate of an organisation, cost efficiency and reduction of costs have always been driving forces in management and are playing an increasingly important role when decisions are taken on access to collections. Many ways can lead to greater cost efficiency and cost reduction. Focussing on physical access, small adjustments in daily routines can help to save time and money. Thinking outside of the box and involving user groups, other stakeholders and staff will help to identify bottlenecks, time-consuming and cost-intensive procedures and will help to develop ideas and ways to tackle them. Who and how many of your staff, from curator to registrar, handle specimens before they are finally sent on loan? Who is involved and who actually needs to be involved during the process of collections access, e.g. researchers, collection managers or even volunteers who assist in collection management? Are there any steps that can be organised more efficiently, e.g. through centralisation or decentralisation within your organisation? Good examples are packing or quarantine routines linked to ingoing and outgoing loans, centralised packing stations and booking systems for visitors to the collections. If internal reorganisation is not the way, what other options are there? Would it be helpful to take external partners or commercial companies on board and to outsource services? What kind of consequences will that have? Is there an opportunity to change collection facilities such as storage systems, cabinets or working spaces to enable easier, quicker and more efficient access while maintaining security measures and conservation standards? New and evolving technologies will open up opportunities for new approaches to collection management. Experiences and advantages of industrial storage systems, such as intelligent transportation and location systems including artificial intelligence, might provide interesting new options and ideas. Why not use them and learn from them, thus bringing together new technologies and responsibility for our cultural and natural heritage?

Exercise

What ways can you think of to reduce costs for access to collections?

What are the "right" decisions to make on access?

Having policies and procedures in place facilitates the routine of quick and easy decision-taking processes for access to collections, from the initial request to the agreement on terms and conditions to the actual visit or loan. In some cases this is not so straight forward. Requests can be very comprehensive when, for instance, many objects are to be sent on loan, much staff time is needed or specific preparation methods are requested. Different requests can be competing for the same resources, for instance when objects are needed for research as well as for an exhibition. Requests might ask for complex destructive sampling or personal involvement and might raise ethical issues.

It is not unusual for staff to continue their scientific and collections work, sometimes for decades, as honorary researchers or volunteers after their retirement. In most cases this is very valuable work and is much appreciated by the organisation. However, there are examples of former curators who, while getting older, did not notice their loss of physical and mental ability to handle objects and do proper scientific work. This might be a very slow process that is masked by the normal ups and downs of performance and may not be noticed and recognised until damage has been done.

How can this sensitive situation be handled? What is the right moment to inform a former colleague that he or she is no longer allowed access to the collection? It is a difficult decision to make, especially when their work has always been appreciated and they are well integrated as colleague, friend, boss, etc. On the other hand, it is your key responsibility to protect your objects from damage or loss.

How do you approach your former colleague when this happens? Is it better to draw a line and prevent access and use of the collections completely at a certain moment which will always be arbitrary? Is it an option to give honorary researchers limited period contracts, e.g. for two or three years which can be extended after yearly evaluation, or is there a way to assist him or her to reduce access step by step? Clear communication will help to avoid misunderstanding, but do realise that there might be a moment in time when communication no longer makes sense; an extremely painful experience for both parties.

Regardless of the level and amount of access requested, it is always wise to thoroughly review a request and to think of alternatives before starting routine procedures. In general, resources needed to enable short-term use should not dominate over the requirements of long-term collection management. Balancing requests from users and demands from your organisation and collections should be the aim. Institutional policies normally indicate what kind of use should be given priority because it supports the mission of your organisation best, e.g. science over exhibition.

Exchange of information with applicants and further discussions on timing or alternative ways can help solve conflicts of interest. What kind of information is really needed? What for? Can it be provided in other ways? Alternatives such as "teaching collections" for student and public training courses are a possible way to prevent damage to valuable scientific objects. Label information, 2D or 3D images, CT scans or other digital surrogates, remote techniques (remote-microscopy) or collection visits are alternatives to physical loans. Emerging technologies will open up even more options. In some cases it might be more efficient to temporarily move entire (sub-) collections to another institution to enable experts to work on them (collection mobility). Applicants often do not know of such alternatives.

Examples indicated on your institutional website will help them to consider these alternatives. Guiding them to online collection databases before applying for a visit or loan could be the first step. Furthermore, it is important to encourage your staff to think of, discuss and develop alternative ways of access internally and with users.

What are the major barriers to access and can you overcome them?

Physical access cannot be unlimited. Resources such as workspaces, laboratory facilities and staff time are limited and apart from that, physical access is restricted by security measures, health and safety issues and conservation requirements. It is important to be aware of the limits of services and to make them transparent as part of your policies and procedures.

> **Exercise**
>
> What is the maximum number of visitors, associates and honorary researchers your organisation can handle per department per day? What are the key factors that limit access to specific collections and services in your organisation?

Access to collections can also be limited by other, in some cases unprofessional barriers. Well known examples include preventing access to specific parts of a collection that someone in your institution is studying or considers "their" collection. Access can also be prevented by ongoing building projects and reconstruction of storage facilities. These are inevitable, often long-lasting annoyances, which do however serve a good purpose. Due to size, ownership, priority setting and historic background, each collection has its own specific barriers.

Barriers to access

A Collections are physically inaccessible:
- Building is inaccessible, e.g. because of building or reconstruction work.
- Storage arrangements do not allow physical access to objects as the collection is poorly organised, e.g. long outdated taxonomy without index.
- Cabinets are inaccessible, e.g. blocked by old furniture, inappropriate storage of collections materials, boxes, etc.
- Objects are physically unavailable because they are on loan, part of an exhibition or extracted for curatorial or scientific work.

- There are no staff available or none with the appropriate competency to provide access to the collections and assist visitors or take care of loans.
- There are staff preventing others, for whatever reason, from studying specific collections under their care or study.
- Health and safety issues restrict or do not allow access to the objects.

B Information or data about objects are missing or incomplete:

- Label or background information is missing or too general and further research would be needed to get this information.

C Collections are digitally inaccessible:

- Collections are insufficiently digitised or the information level is too low.
- Data are not available from the outside (online-portals) but only on local servers/computers.
- Intellectual Property Rights (IPR), General Data Protection Regulation (GDPR) or other legal issues (ABS, CITES) prevent data and information from being accessed.

Barriers and bottlenecks slow down processes, making them inefficient, costly and time consuming. Sometimes they restrict or even prevent access to collections and consequently the use of them.

Exercise

What are the biggest barriers for physical access in your organisation and collections? What are the main factors? What are the mechanisms behind them? Will a collection assessment help you to identify bottlenecks? Do you have a strategy on how to overcome these barriers? What are possible steps to reduce them to enable and facilitate access? What can be organised more efficiently?

Strategies to overcome barriers

Decisions on whatever level should have a solid background. A clear vision helps to develop a strategy and set priorities. Often more than one step will be necessary to reach the final goal. As a first step, make sure that your collection is physically accessible upon request. Knowledgeable staff, unique identifiers and a digital index will help you. If comprehensive digital access is not achievable in the short term, start with a basic set of information on higher taxonomic groups or collection units. Providing this information in online databases and on the website will help answer requests from users and collection management. The online presentation of high resolution scans of whole insect drawers at the Museum für Naturkunde Berlin is a good example of this (www.museumfuernaturkunde.berlin/en/science/eos-establishing-inventory-specimen-rich-collections). They give an overview of an entire collection management unit. First results of this experiment show that requests are much more focussed. Internally, the images support the improvement of collections through a better priority setting process.

What will access look like in the future and how can we be prepared?

The way collections are accessed and used is changing over time. For natural science collections, research trends and new technologies have been the driving forces. Digital transformation influences the way research is conducted and how collections and related information are needed and used. User requests are shifting from access to physical objects to requests for data and media such as 3D images, CT scans, etc. New technologies enable new ways of research that answer new kinds of questions but are still focussed on the same material (technology push). Proven by successful examples, collections data, information and expertise can be useful for many communities. Users are no longer solely scientists from core research fields, such as evolution or biodiversity discovery, but include increasing numbers of amateurs, citizen scientists, and the interested public (market pull). It is now one of the major challenges of organisations to open up their collections and expertise for a new audience in order to make themselves more relevant (see Chapter 2).

Organisations need to change their perspective and way of providing access, evolving from an internal focus on core user communities to an outward looking, customer-oriented approach. That has clear effects on the portfolio of tasks and services. We cannot foresee the future, but we can try to be prepared as best as we can. Trends we are already observing can serve as starting points for considerations. Market analyses and cost-benefit relations will help in developing strategies and setting priorities. What kind of access, data and services are requested and by whom? Are there other user communities that could benefit? Is there a return on investment? What is the benefit for the organisation of opening up the collections for a new kind of use and related user community? Natural science collections are mostly ordered according to taxonomy, whether in alphabetical sequences of genera and species or by systematic treatments. This collection arrangement facilitates access for colleagues from the same user community (taxonomists). New users and uses might need different subsets and different approaches to answer their questions, e.g. interdisciplinary work that involves people from a different background and with questions that need a geographic or periodic arrangement and brings together objects and parts of the collections that are organised and stored in different systems. Virtual collections can help users to select and combine appropriate data and information. However, in some cases, new methods of collection management, and maybe even physical arrangements, might be needed. The same is true for new analytical methods that ask for different handling and access and even produce new subsets of the collections, such as microscopic slides or sections that need to be stored in a different way. See Chapter 2 which describes past, current and future uses of collections.

Digital transformation

New technologies are changing the ways of living, working and research, including a shift from physical to digital. This also applies to collections. Instead of applying for physical access there is an increasing demand for digital information. In the 1990s, natural science collections started to digitise basic collection data (what, from where, when and by whom). Soon, self-made databases became complex collection data systems developed further into collection management systems handling collection data, media and management processes that were opened up online. This development is still going on. Information from collections is getting more and more linked with other sources such as publications, historic and current events, people, etc.

A survey conducted during the EU project SYNTHESYS in 2017 concluded that online data and information on collections do not necessarily reduce the number of requests for physical access (see www.synthesys.info; Report on effects of digitization on physical access and loans). For individual organisations this might well have an impact. At the Smithsonian National Museum for Natural History in Washington, images of botanical type specimens reduced the number of loans of types by 80%. Online data and information helps to prepare and focus requests for physical access and reduces the workload for collections management staff.

Online accessibility also increases the visibility of your collections. It raises awareness for collections and invites new people and potential users to get connected. It is now up to the community of natural science collections to facilitate this. A first step is to link data and information and provide them through joint databases and portals such as GBIF. A second step is to adapt management systems and develop mechanisms that allow requests for access to objects distributed over several organisations through a single entrance point, as is planned through DiSSCo (Distributed System of Scientific Collections; see www.dissco.eu).

The growing amount of online collections raises new questions and challenges: What will happen with data and information and how will we ensure that the connection to the collection objects and the institution of origin will be maintained? Is it within or beyond our influence what external users will do with the data? How do we reach out to and encourage them to link back to the organisation of origin (see Chapter 7)? Users, especially those who use our digital data without being known to us, become a challenge for the future. How do we reach them, how do we make them part of our collection management and information system to further enrich our collections?

Summary

Access to and use of collections is key to maintaining their relevance. Value of natural science objects particularly derives from their use and ongoing contextualisation by connecting them with information, placing them in another, new context that then builds the basis for new questions and use. Access and use of collections can be seen as the beginning of a value chain, from object to knowledge and impact. Encouraging and facilitating access to a broad range of uses and user groups should be in the core interest of any organisation holding collections.

At the same time, we are responsible and accountable for the protection and preservation of the objects we are managing. Therefore, enabling access while minimising risks to the objects is the overall guiding principle. The framework is set by (inter)national rules and regulations, institutional priorities, by conditions such as staff resources and structure, and the accessibility of the collections. Access can be given to objects through loans or visits (physical access) and to (meta-)data and information linked to them (digital access). General principles help to deal with different types of uses regardless of the organisational structure. Institutional policies and procedures should be clear about whether and how access can be given to objects and data, about processes, criteria and responsibilities. Transparency and a proper documentation of all processes are key factors for efficiency of internal processes, clarity for users and a safety factor in case of misuse, theft or damage of objects.

Number of visitors, loans, samples, requests, downloads, etc. are the most common performance indicators for access. Whenever possible, analyses of purpose and context of access and use should be added to improve your understanding of the strengths of your collection and potential impact and relevance for science and society. All analyses and documentation must be legal and ethical and in accordance with the General Data Protection Regulation.

Access is not gratis but incurs costs, e.g. expenses for packing, mailing and external services, staff time in collection management and administration, maintenance costs and others. Limits are given by workspaces and staff time and by security measures, health and safety issues and conservation requirements. Making processes more effective and efficient, and thus reducing costs, has always been a driving force and plays an increasing role in the decision making process. This can range from small adjustments in daily routines, the implementation of new methods and technologies to more strategic approaches and reorganisations. In any case, your institution always retains the right to deny, restrict or withdraw access to the collection, e.g. in the case of failure to comply with legal or ethical regulations or exceptional high risk to the objects.

Driven by research trends and new technologies, and more recently by societal challenges, access and use of the collection have always been changing over time. Currently we are observing a shift from requests for access to physical objects to requests for data and media such as CT scans, 3D images, etc. Users are no longer solely researchers but include an increasing number of amateurs, citizen scientists and the interested public. We cannot foresee the future, but we can try to be prepared as best we can. That includes a change in our perspective, from the strong internal focus on our core research community to an outward looking, customer-oriented organisation. That has clear effects on the portfolio of tasks, services and priorities. Enabling barrier-free digital access (easy, timely, convenient, customised) is one of the biggest challenges.

Further reading

Abrahamson, B.L., 2015, Tracking Changes In Natural History Collections Utilization: A Case Study at the Museum of Southwestern Biology at the University of New Mexico. *Collection Forum*, 29: 1–21.

Carter, D.J. and Walker, A.K., 1999, Policies and Procedures. In: D. Carter and A. Walker (eds.), *Care and Conservation of Natural History Collections* (pp. 177–192). Oxford: Butterworth Heinemann.

Collections Trust. Spectrum UK Collection Management Standard. https://collectionstrust.org.uk/spectrum/ [Accessed 22/0719]

Dorfman, E. (ed.), 2018, *The Future of Natural History Museums (ICOM Advances in Museum Research)*. London and New York: Routledge.

Pettitt, C., 1994, Manual of Natural History Curatorship. In: G. Stansfield, J. Mathias, and R. Gordon (eds.), *Using Natural History Collections* (pp. 144–166). London: HMSO.

Simmons, J.E., 2006, Access and Use. In: J.E. Simmons, *Things Great And Small: Collection Management Policies* (pp. 111–115). Washington, DC: American Association of Museums.

Examples of policies: visits, loans, destructive sampling

The Museum of Vertebrate Zoology, University of California, Berkeley: mvz.berkeley.edu/Specimen_Loans.html

Florida Museum of Natural History, Gainesville: www.floridamuseum.ufl.edu/collections/policy/

Naturalis Biodiversity Center, Leiden: science.naturalis.nl/en/collection/loans/

Natural History Museum, London: www.nhm.ac.uk/our-science/collections/accessing-collections.html

American Museum of Natural History, New York: www.amnh.org/research/sicg/amcc/collections-policies/use-and-care-of-the-collections
Smithsonian National Museum of Natural History, Washington: naturalhistory.si.edu/research/botany/collections-access/collections-management-policy

APPENDIX 8.1: PRINCIPLES FOR RESEARCH LOANS BETWEEN NATURAL HISTORY COLLECTIONS

The document reproduced here was created by the Directors of Collections group of the European Distributed Institute of Taxonomy (EDIT). EDIT was funded as a "network of excellence" (NoE) by the European Commission (6th framework programme, FP6) and we would kindly ask that this is acknowledged in any re-use, re-purposing or dissemination of these materials. This report reflects the views only of the authors. The Commission cannot be held responsible for any use which may be made of the information contained therein.

CPB principles for research loans between natural history collections

Introduction

The following document was drafted and revised by the EDIT[1] Directors of Collections (DoC) group. It is based on the loan policies of 13 European natural history museums[2] participating within the EDIT project, prepared by a taskforce within the DoC, and the result of discussions held during DoC workshops and feedback from DoC member institutions on previous versions. Thirteen EDIT institutions signed the loan policy document for the duration of the project.

With the end of the EDIT project, the DoC continued as the Collections Policy Board (CPB) under the umbrella of CETAF. New partner institutions joined the CPB. At their first meeting in Madrid in September 2011, the loans principles document was evaluated. It was generally agreed that this document could, without alteration, be adopted by the CPB institutions.

Aim

On the basis of international agreements and regulations and general collections policies, the document aims at common principles for research loans within CPB institutions that will facilitate access to collection material through loans whilst maximising their long-term preservation. This is closely aligned with the CPB mission to develop commonality of policy and implement common standards and procedures in collections management as outlined in the CPB Business Plan 2011–2014.

This document covers the loan of both morphological specimens and molecular collections such as frozen tissue and DNA for scientific purposes. It does not cover living collections, libraries and archives, loans for exhibition or loans for commercial purposes.

Background

Loaning specimens or samples whether morphological or molecular has a number of clear benefits to both present and future users of collections:

- allows users to compare material;
- adds value to collections through annotation;
- avoids unnecessary travel.

However movement of loans inevitably puts collections at risk. These risks include:

- damage or loss through poor handling in post;
- confiscation due to failure to comply with changing laws on material transfer, etc.;
- damage due to irradiation, etc. on entry to country;
- "professional competition", i.e. individuals borrowing material to prevent others using it;
- damage or loss through lack of clear institutional responsibility.

Policies are designed to minimise these risks while facilitating responsible access. Inevitably through tradition and legal requirements, collections policies vary both between and within institutions and in some cases there are no policies. The CPB agrees that there is an urgent need to improve and harmonise loan procedures in order to:

- prevent the loss of and damage to collections;
- guarantee transparency including traceability of collection items, i.e. where are they, who is responsible for them, and what are they being used for;
- ensure that collections on loan will be managed to the same quality standards and rules in all institutions;
- encourage alternatives to loans where possible such as:
    - digital images, remote microscopy, etc.;
    - subsidised collection visit (e.g. through SYNTHESYS);
    - label information through databases.

Definition of terms

Institution

In this context "institution" refers to a museum, botanical garden or university department holding natural history collections that has a common governance. For instance a geological museum, a zoological museum and a herbarium within the same university would be considered as three institutions unless they had a clear, common, overarching management structure for their collections.

Material vs. specimen

Material is a collective term for a set of specimens and any other associated items sent on loan. Specimen relates to a single unit of biological or geological origin with associated data such as provenance, collecting date, collector, etc.

Principle

In this context a principle is an action or way of working that the signatories agree to continue or put in place within an agreed time period.

General policy statements

As a matter of general collections policy the signatory institution is committed to the following:

1 Endeavouring to follow relevant national ethical codes for museums with reference where possible to the ICOM Code of Ethics for Museums.
2 Abiding by all international and national agreements governing the transfer of biodiversity specimens and products such as CBD, CITES and other agreements on access and benefit sharing, e.g. the Bonn Guidelines[3].
3 Having collections management policies that will meet the requirements of any agreed European or national accreditation standards wherever possible.
4 Having all systems for storage, managing and archiving information relating to loan transactions that will comply with agreed standards for the documentation of items, e.g. Spectrum.
5 Complying, where necessary, with dangerous goods regulations (i.e. IATA) governing specialised transportation of material.

Key principles

In line with the general principle of maximising access to our collections the institution agrees to the following:

1 All specimens are available for research loan.
   - However, in any circumstances the institution reserves the right to refuse to lend any material within its collection at its discretion. The reasons for refusal should be transparent and may include excessive cost and unacceptable risk to items such as type and figured

specimens, extinct (recent) species, specimens of high historical significance, and delicate or hazardous materials, etc.

2 There will be no charge for research loans, e.g. packing materials, postage and staff for the signatories.
   - Charges may be made for repairs to or replacement of damaged specimens, for specialist packing, treatment or for any unusual or exceptional means of carrying.

3 The institution in which the loan (or material associated with the loan, e.g. documentation) is to be housed and used must in all respects be safe, secure and conform to the recommendations and requirements of the lending institution.

4 Material sent on loan will only be used for research, not for commercial purposes without prior agreement.

5 The borrowing institution accepts that title to and ownership of the loaned items remains with the lending institution at all times.

Morphological collections

Material available for loan

- Only an agreed proportion of the institution's holdings of a collection unit, e.g. species or samples from one locality, will normally be sent on loan at any one time.

Loan requests

- Material will be sent to (non-profit) institutions with natural history collections and staff responsible for collections management (including universities, if they can comply with the conditions within this document). In exceptional circumstances loans will be made to individuals at private addresses, but then only under stringent additional terms and conditions as defined by the lending institution.
- Only approved borrowers[4] can borrow specimens.
- Responsibility for the borrowed specimens rests with the institution in which they are kept or in exceptional cases with individuals at private addresses (see above).
- Requests for loans should state the names of the researchers on whose behalf they are made by the borrower, together with the scope of the work contemplated, and intention of publication.
- It is expected that all signatories will work towards the following standards for management of loans within an agreed time frame:
   - Loan requests are preferably submitted to a central institutional or departmental web address or as an interim to (a) designated person(s) as the lending institution defines.

- As much information as possible should be given when requesting items for loan, e.g. registration numbers and synonymy.
- Details of all incoming loans must be recorded electronically to enable tracking all material entering as loans centrally or at least at department level through a single contact point.
- Forwarding of loans to third parties is not allowed without prior written approval of the lending institution[5].

The loan term and loan extensions

- Normally, the maximum initial period for research loans is 12 months. If approved for extension, the renewal period will be for 6 to 12 months with the possibility to extend twice; the maximum loan duration is therefore 3 years.
- In some circumstances research loans will be approved for an initial period of up to 3 years, e.g. for comprehensive revisions or PhD studies, when sufficient explanation for such a duration is given. Such loans can be extended for another 2 years up to a maximum loan duration of 5 years.
- When applying for an extension, the reason behind this extension has to be made clear in the application, including a scientific progress report on the work conducted so far and the work that still has to be done.
- Application for extension must be submitted well in advance, at least a month before the end of the current loan agreement.

Photos and copyrights

- Taking and publishing photos of material on loan is only allowed after written approval[6] by the lending institution.
- Reference to photos of specimens on loan must be forwarded to the lending institution and linked to the objects, e.g. in separate letter or in a collection database according to guidelines provided by lending institution.
- Photos will become available to the owning institution.

Restrictions

- No changes to original labels are allowed.
- No changes to specimens or any other material are allowed without permission (including pedestals/armatures, sheets, etc. to which the specimen is fixed).
- DNA or tissue sampling from specimens on loan is only allowed after approval by the lending institution. If approved, the principles for molecular collections apply (see below).

- Annotations by the borrower should not be permanently attached but provided otherwise, preferably electronically or by determination slips, labels or other hard copy means.
- If an object is damaged, the owner must be informed immediately, no repairs are allowed without consultation with the owning institution.

Non-DNA destructive sampling, e.g. dissections of insect genitalia, flowers, geological samples, etc.

- Non-DNA destructive sampling by the borrower is only allowed after approval by the lending institution.
- Unused fragments of specimens, etc. are the property of the owner and must be returned in an appropriate capsule, etc. and clearly labelled.
- If destructive sampling leads to other products (such as slides or geological objects) and the borrower is allowed to keep products, the institute of origin has the right to ask for duplicates.
- When there are multiple requests from different borrowers for destructive sampling of the same material only one application may be approved, but all applicants will be informed and encouraged to collaborate.

Acknowledgements and feedback

- Annotations or data of collection material made digitally available by the borrower should become available to the lending institution.
- The lending institution must be acknowledged using the official name in publications arising from research on their material.
- Reference to specimens in publications should where possible be through their unique identifiers such as registration numbers, collectors numbers, etc.
- Publications arising from research on borrowed material are to be sent to the lending institution (preferably as PDF).

Housing, security, environmental conditions, packaging and transport

- Loans will only be sent to institutions who have specified the way the borrowed specimens will be disinfested upon receipt. This report is needed only once and will be applicable to all future loans unless changes to disinfection are made. The protocol must be acceptable to the lending institution.
- Preferably, loans are to be returned in the original packing.
- In exceptional circumstances, e.g. fragile specimens, the institutes involved must agree beforehand on the means of transport, but it is the owner who decides. The lending institution can request to return the

loan by an official carrier company, by registered mail or, e.g., personal transport.

- Confirmation that the loan has been received and in good order has to be sent upon receipt. The borrower also has to inform the lending institution when the loan is to be sent back.

Failure to comply with the terms and conditions imposed

- In the case of failure to comply, the loaning institution shall contact (in writing) the borrower and the borrowing institution and request that the situation is rectified immediately.
- Following the above, if the approved borrower still fails to comply with this written request, the lending institution shall be entitled to terminate the loan and recover the loaned items immediately.
- If the lending institution is forced to terminate (temporarily) the loan it may apply sanctions to future interactions with both the borrower and the borrowing institution.

Molecular collections

Molecular collections (see Appendix for definition) form a discrete group of collections not only in the technical expertise required to set them up, manage and develop them but also in their housing. Specific statements are included below that apply to these collections.

Loan requests

It is expected that all signatories will work towards the establishment of a coordinating mechanism for molecular loans, e.g. a "Specialist Molecular Curator" or "Molecular Advisory Panel" to:

- advise and confirm feasibility of success for the destructive sampling loan request for molecular work (as supported by evidence of past successes for the molecular processes intended on application form);
- assess specimen/sample status to minimise damage, over-use of sample and impartially prioritise loans in the case of multiple identical requests, especially from type, extinct or rare specimens, informing and inviting collaboration between all researchers involved;
- review exceptional circumstances, e.g. a request which will reduce the original sample to nothing remaining;
- suggest the use of the on-site facilities of the institution of origin (molecular laboratories and sequencing facilities) for either visitors or their own staff to perform the destructive molecular sampling if the specimen or sample is too precious, fragile, sensitive (rare/endangered, CITES, etc.) to send away.

Obligations of the applicant for molecular loans

- All remaining samples from the original loan material for destructive sampling must be returned: DNA/RNA/protein/metabolite/lipid, etc. extractions, and if requested PCR products, sequencing reactions, and direct copies of sample data including sequence data. If it is not possible to return these items the applicant must inform the lending institution.
- A storage compatibility check guarantee in writing is required for any molecular collections to ensure that appropriate matching storage conditions are used in the borrowing institution.

Acknowledgements and feedback for molecular collections and products

- The applicant must notify the relevant department of publications arising from the destructive sampling of the specimens or samples and will be asked to provide copies including acknowledgement of institution of origin in all subsequent publications, GenBank numbers, etc.
- In the case of DNA studies the institute of origin must be informed of any data sent to GenBank, e.g. object accession number.
- Any data sent to GenBank should be linked to the original specimen and accession or similar unique identifier.
- Reporting the results to the owner is compulsory, including negative (no) results.

Intellectual property rights/property rights for molecular collections

- The Intellectual Property Rights (IPR) of original genomic DNA, RNA, protein extractions, progeny and non-modified derivatives lies with the institute of origin; hence forwarding any part of these to third parties will require prior permission from the lending institution and mutually agreed terms, as defined by a material transfer agreement (MTA).
- If samples are sent to the applicant, the institution's research loan policy and any associated loan terms and conditions apply. The applicant will sign and return a receipt to acknowledge the arrival of the material and agree to the terms and conditions.

Warranties for molecular loans

- The lending institute is not liable for failures in any molecular analysis (DNA extraction, PCR product, sequencing reaction, etc.).
- The lending institute makes no warranties as to the safety of the material; the material is experimental in nature and must not be treated as if it is free from contamination. The recipient assumes full responsibility

for complying with the recipient nation's quarantine and bio-safety regulations and rules as to import or release of genetic material. However, loan shipments from the lending institute containing known contaminants and/or biohazards will be accompanied by paper declarations and supporting best practice documents for molecular work in the light of these contaminants.

Collections Policy Board December 2011

Appendix: Molecular collections

Definition

Molecular collections/samples are intended specifically for molecular research

This includes:

Samples/Specimens

- deep-frozen specimens or tissues in liquid nitrogen (-196°C);
- frozen tissues in freezers (-80°C);
- alcohol-preserved specimens specifically collected for molecular work or tissues (-20, +4, +16°C);
- silica-dried material (RT);
- lyophilised/freeze-dried material (RT);
- samples on FTA cards (RT);
- viable cell lines cultured from fresh tissues (cryo banks).

Products

- extracted genomic/mitochondrial/chloroplast DNA in water or buffers (-20°C);
- extracted RNA (-80°C);
- genomic/expression (cDNA) phage/BAC libraries in bacterial culture (-80°C);
- expression (cDNA) and genomic libraries on FTA cards (RT);
- DNA Sequencing reactions;
- PCR products;
- protein and amino acids;
- Mixed Environmental Genomic DNA/RNA/protein and amino acid survey samples;
- historical contaminant sample series;
- lipid and metabolite sample series;
- returned material from all destructive sampling loans;
- unforeseen future analyses.

APPENDIX 8.2: PRINCIPLES FOR ACCESS BY PERSONAL VISIT TO NATURAL HISTORY COLLECTIONS

The document reproduced here was created by the Directors of Collections group of the European Distributed Institute of Taxonomy (EDIT). EDIT was funded as a "network of excellence" (NoE) by the European Commission (6th framework programme, FP6) and we would kindly ask that this is acknowledged in any re-use, re-purposing or dissemination of these materials. This report reflects the views only of the authors. The Commission cannot be held responsible for any use which may be made of the information contained therein. This document was developed under the auspices of the Consortium for European Taxonomic Facilities.

Collection Policy Board[7] Principles for access by personal visit to natural history collections

Introduction

The following document is based on the visitor access policies of several European natural history museums which are members of the CPB. It was prepared by a task force of CPB and has been modified and revised as the result of discussions and feedback following a CPB general meeting in Paris in September 2012.

Aim

This document aims to provide a set of common principles for facilitating access to CETAF collections through personal visits whilst minimising risks to their long term preservation. This is closely aligned with the CPB mission to develop commonality of policy and implement common standards and procedures in collections management as outlined in the CPB Business Plan 2011–2014. This document covers visits to morphological collections and related material such as archives. It does not cover visitor access to molecular collections such as frozen tissues and DNA, living collections and libraries.

Background

Access to collections by personal visit has clear benefits to both present and future users of collections and it is welcomed and encouraged by CETAF institutions. Visits:

- allow users to examine specimens that for reasons of fragility, rarity, sensitivity and/or international regulations cannot be sent on loan;

- add value to collections through annotation;
- allow and encourage discussion with research and collections staff.

However visits might also expose collections to risks such as:

- damage or loss through poor handling;
- loss through vandalism or theft.

Inevitably through tradition and legal requirements, collections policies vary both between and within institutions and in some cases there are no policies.

The CPB agrees that there is an urgent need to improve and harmonise access procedures taking into account new developments in collection access such as digital images and remote microscopy.

These principles are designed to minimise risks while facilitating responsible access.

Definition of terms

Institution

In this context "institution" refers to a museum, botanical garden or university department holding morphological natural history collections that has a common governance. For instance a geological museum, a zoological museum and a herbarium within the same university would be considered as three institutions unless they had a clear, common, overarching management structure for their collections.

Principle

In this context a principle is an action or way of working that the signatories agree to continue or put in place within an agreed time period.

Legal framework

As a matter of general collections policy the signatory institution is committed to the following:

- endeavouring to follow relevant national ethical codes for museums with reference where possible to the ICOM Code of Ethics for Museums;
- abiding by CBD and other international agreements on access and benefit sharing, e.g. the Nagoya Protocol;
- having collections management policies that will meet the requirements of any agreed European or national accreditation standards wherever possible;
- complying with appropriate health and safety regulations to ensure safety of visitors and staff.

Key principles

In line with the general principle of facilitating access to our collections to a greater diversity of people, the institution agrees to the following:

1 To make its collections available to *bone fide* scholars and researchers in any discipline, providing that:

- the specimen is not on loan, or out of the collection for any other reason;
- the specimen is not culturally sensitive, unusually fragile, light sensitive, at a high risk of theft;
- the security of the staff and collections is not compromised;
- access does not unduly interfere with any programmed curatorial or research work.

2 To assist the visitor through expertise and technical advice.
3 To endeavour to provide access to the information relating to the collections unless it is deemed that the information should be kept confidential.

Governance

Access to research and reference collections is controlled by an appropriate nominated senior officer as designated by the institution.

Access and security

Visitors must book visits in advance stating what they wish to see and for what reasons. Institutions are within their rights to cancel and/or deny visits such as unannounced visits, emergency situations or unforeseen circumstances.

Institutions should have visitor instructions in place. Visitors should be informed about them beforehand.

All visitors should sign a visitor book or form including name, address, institution if appropriate, collection visited and date of the visit.

Bench fees/Charges

Charges for visits may be made by some institutions in certain cases. If charges are likely to be made then the extent and reason for these must be made clear to the visitor before the visit.

Valuations

Staff do not provide visitors with valuations in financial terms for any objects unless they are duly authorised or requested to do so by the institution.

Photography and copyright

Visitors to the collections are allowed to take photographs but these will be subject to approval of the collections management staff and copyright agreements.

Restrictions

No changes to original labels are allowed. Any annotations by the borrower should not be permanently attached but provided by other means, preferably electronically or by determination slips, labels or other hard copy.

No changes to specimens or any other material are allowed without permission including DNA or tissue sampling, dissections, pedestals/armatures, sheets, etc. to which the specimen is fixed.

Collections Policy Board 11th December 2012

APPENDIX 8.3: COMMON PRINCIPLES ON DIGITAL DATA PRODUCED BY EXTERNAL USERS OF NATURAL HISTORY COLLECTIONS

The document reproduced here was created by the Collections Policy Board, a working group under the umbrella of the Consortium of European Taxonomic Facilities (CETAF). The work was conducted through the project SYNTHESYS3 (Synthesis of systematic resources, 2013-2017). SYNTHESYS3 was funded by the European Commission in the 7th framework programme (FP7-INFRASTRUCTURES, specific programme "Capacities": Research infrastructures) under the theme INFRA-2012-1.1.16. - Natural History Collections. Would kindly ask that this is acknowledged in any re-use, re-purposing or dissemination of these materials. This report reflects the views only of the authors. The Commission cannot be held responsible for any use which may be made of the information contained therein.

Common principles on digital data produced by external users of NH collections (final version 15th December 2016)

Introduction

The present document was created within the framework of SYNTHESYS3 NA2 Objective 1 Task 1.1 / "Develop policies for virtual CM and integrate JRA outputs", which is led by the Museum für Naturkunde, Berlin (MfN), with the participation of the following institutions: NHM, RBGK, RBGE, UCPH, CSIC, RMCA, NHMW and NMP. It is a result of research on available digital data management policies as well as other principles and adapted to the requirements of European institutions housing natural history (NH) collections based on discussion with project partners and CPB[8] members.

Aim

On the basis of international agreements and regulations as well as general collection policies, this document aims at common principles for the management of digital data produced by external users of NH collections.

The policy statement aims to help maintaining the highest professional standards for the management of digital collection data by museums and other institutions with digital collections. It creates a framework for the management and conservation of digital data produced by external users of NH collections hosted by European institutions. It clarifies authorities and responsibilities and guides the necessary workflows for implementation.

Background

Gathering and providing digital collection data has a number of clear benefits for both users and owners of collections:

- support of the physical specimen and its management by its associated digital data, e.g. by providing additional information to the object or as a backup;
- preservation of national and international heritage and prevention of loss of (physical) objects, knowledge and connections due to minimisation of object handling and shipping;
- facilitating or even enabling global and efficient research by facilitating broad access to the collection, focussing, reducing and increasing the efficiency of physical visits to the collection or by minimising the object handling and the use of microscopes and other magnification equipment.

External users often have excellent knowledge on different aspects concerning NH collections (e.g. on taxonomy, localities, etc.) and could help to utilise and update information on collection data and material. However, this information is not always made available to the institution by external users and the institution does not profit from this valuable source of information. In other cases information is left as notes on paper somewhere in the collection which causes more work to limited institutional staff and errors may occur in the process of transcription.

Hence, it would be a great advantage if external users of NH collections would contribute directly to the collection database. However, this is often not feasible due to the complexity of these databases and due to the fact that data need to be validated before they are entered into the database.

Definition of terms

Digital data

Digital data of concern in this policy statement are data associated with collection material. They may derive from physical or digital objects held in the institution or from the collection itself. They may comprise of scientific raw data as well as data as products of a creative act in the form of, e.g., texts, digital images, transcriptions, metadata (such as statistics) or datasets. Digital data in this document refers to raw data and images derived from the objects held by the institution and not secondary data produced by the user.

External user (here also: "user")

External users of NH collections may be visitors and guests of the collection as well as borrowers of collection material of an institution. Users will be asked to comply with terms and conditions of access by the hosting institution.

Institution

In the present context, the term "institution" refers to a museum, university department or similar holding a NH collection under one governing body/authority. The institution referred to in this policy statement holds the collection or specimen the digital data of concern are linked to.

Principle

In the present context a principle is an action or way of working that the signatories agree to establish and implement within an agreed time period.

Policy statement

Declared objective that an institution aims to achieve and establish. It is envisaged that institutions will implement this objective through their own policies and governance.

Specimen

Specimen relates to a single unit of biological or geological origin with associated data such as locality, collecting date, collector, etc.

Legal Framework

The signatory institution is committed to the following:

- follow relevant national ethical codes for museums;
- endeavour to follow the ICOM code of ethics for museums;
- abide by all relevant international and national agreements governing collections;
- have management policies on digital collections in place that meet requirements of any agreed European or national accreditation standards where possible;
- follow the Bouchout Declaration;
- observe and implement relevant international agreements such as Convention on International Trade in Endangered Species of Wild Flora and Fauna (CITES), Convention on Biological Diversity and the Nagoya Protocol on access and benefit sharing.

General principles for digital collection data

The institution is committed to:

- valuing the contribution made to the collections by external users;
- having systems for long-term storage and maintenance of digital collection data in place;
- setting quality standards for digital collection data (in terms of, e.g., resolution, available metadata and specimen data, unambiguous linkage of data to object);
- observing any legal or contractual requirements relevant to the collections accessed and data gathered;
- ensuring the safety, longevity and long term readability of stored data.

Key principles for digital collection data produced by users

To accomplish the general principle of optimising management of NH digital collections the herewith undersigned institution agrees to the following:

1 Access to collections for external users will be facilitated where possible and encouraged.

2 All digital data as specified above and annotations of collections produced by external users should be made available to and usable by the institution owning the specimen.
3 Each institution decides which data it intends to keep and to make available.
4 The institution supports Open Access taking "good scientific practice" into account. This can mean that data are kept by the institution with an embargo of a specific time on the raw data, e.g. three years to allow researchers to publish the results first before the data are published open access.
5 Users should be informed of data the host institution intends to gather and publish.
6 All necessary forms as well as paperwork will be provided in English as well as in the institutional native language.
7 All relevant forms and paperwork as well as the institutional Management Policy on Digital Collections will be made easily accessible.
8 The user should be asked how they intend to use the data gathered. This information can be used to assist the institution in refining its practices to meet user requirements.
9 The institution has clear and comprehensible procedures established to deal with digital collection data produced by users, including responsible persons to contact and submission protocols (e.g. submission through a centralised management).
10 Digital data should be provided in a standard format that ideally can be fed directly into the institutions database system (e.g. institutional standard file formats, standard field and data structure for datasets).
11 Producing digital material such as photos, films, etc. of collection material by users under conditions of use should conform to institutional policies.
12 The user should clearly cite and acknowledge the institution in all outputs and publications based on or including digital data to enable discoverability. This should use the institutions preferred citation structure (e.g. the CETAF Stable Identifier) which should be included in papers as well as submissions to aggregators and online databases (e.g. GBIF, Genbank, BOLD and Morphbank).

Notes

1 European Distributed Institute of Taxonomy; EU project 2006–2011. The DoC group was set up in 2008.
2 MNHN Paris, NHM London, NNM Leiden, ZMA Amsterdam, MfN Berlin, NBG Brussels, RBINS Brussels, RBG Kew, IPAN Krakow, NHN Leiden, SMN Stuttgart, RMCA Tervuren, BGBM Berlin.
3 See www.cbd.int/abs/bonn.shtml

4 Approved borrowers must be bona fide researchers; in case of postgraduate students, postdoctoral researchers, emeritus and retired staff, honorary or adjunct staff, etc. requests must be made through their supervisor, mentor or responsible staff in charge, preferably a single person of position such as the head of collection management.
5 The approved borrower undertakes to ensure that the loaned items shall only be kept at their host institution unless specific permission has been granted by the appropriate department to allow the item to be moved to another locality.
6 Approval to the borrower to make photographs is not same as transfer of copyright. Copyright must therefore be part of loan agreement. The photographer (if it is the borrower) has to be mentioned in case of publication of illustrations.
7 Collection Policy Board (CPB) is a working group of Directors or Heads of Collections under the umbrella of Consortium of European Taxonomic Facilities (CETAF).
8 Collection Policy Board under the umbrella of CETAF.

9 Asking the right questions – putting it all into practice

Introduction

Imagine, one day you are appointed as Head of Collections or Director of a museum responsible for staff, budget, collections and/or exhibits. In the real world, this will not happen suddenly. However, not many of us will be prepared for such a promotion because helpful guidance can be difficult to find. What is important is to be able to ask the right questions at the right time and get the answers you need to move forward. In this book, we have tried to give you an overview of some of the issues that you will have to deal with and to provide you with ideas on how to tackle them. Now it is time to practice what you have read and learned: identify priority areas for action, ask the right questions, develop a strategic approach and provide leadership.

Below you will find a series of questions. They serve as guidance through a period of starting up and planning, or as one of us stated "I wish I had this available during my first 100 days in office". They are not meant to be exhaustive or as a project management handbook but as a stimulus to trigger your thinking. Read them. Do they apply to your situation? Did you ask yourself similar questions? Are you able to answer them or do you need further advice or information to be able to do so? If they do not apply to your specific situation, what questions are missing? Try to formulate and, of course, answer them. Enjoy it and good luck!

Scenarios

Before you start, it is important to have a clear understanding of your position and role. Take a few minutes and describe your position, responsibilities, expectations from your manager and staff, the strategic framework of the organisation and the challenges ahead of you. The clearer you can paint this picture the better you will understand your role in the system and the easier it will be to identify priority areas for action.

It might also be that you are preparing for a new job and cannot yet describe your future position or that you want to practice it more extensively.

In these cases, you can create a scenario based upon your own experiences and ideas or by using one of our three sample scenarios below.

Scenario A: molecular collection

A museum or university with traditional biological collections is aware that it also houses specimens and DNA samples collected for molecular analysis. These are potentially valuable resources and should be regarded in the same way as other collections. There is a directive from the director to improve these collections and make them more accessible. Currently they are distributed in refrigerators and cupboards in multiple locations. They are believed to be mainly insects, vertebrate tissues, blood samples and some feathers. A project plan was formulated and additional funds were acquired to realise the project.

You have been appointed as the project manager. Within the framework of the project (description of staff, time and budget), you are asked to develop a "cold archive" that fulfils the requirements of a modern research infrastructure (standards for storage, access and use). You report directly to the director. The support of all staff involved, especially the scientists, is essential for the success of the project. What would your first 100 days look like? Where would you start and why? What pitfalls do you expect? How would you prepare yourself and your team for this project?

Scenario B: digital collection

You have recently been appointed as director of a small multidisciplinary museum with valuable natural science collections. Staff and budget are limited and due to the interests of your predecessor, the budget was mainly spent on the cultural aspects of the museum. There are a number of humanities objects digitised and web accessible but no natural science objects. In order to make your collection, and hence your museum, more visible and connect it to the scientific community, there is a need to make the collections, especially natural science, more accessible digitally and to contribute to data aggregators such as GBIF. How would you approach this challenge? Where would you start and why? Which questions need to be answered? What would you do if you did not have the necessary expertise?

Scenario C: cultural collection

You are an established senior curator in a medium sized museum, and you intend to make improvements in the care of the museum's cultural collections. The collection includes items from several different religious traditions, works of art, items that reflect every day activities and human remains. Your project includes moving the cultural collection into a new facility with new storage equipment, possibly according to a new arrangement. You see the move as the opportunity to verify the inventory, update and digitise the early 20th century catalogue records, including making digital photographs for record-keeping purposes. You have heard that some museums follow culturally sensitive practices when caring for objects from a different culture, also known as "traditional care", but do not know how to apply it. Is traditional care relevant in your context? How would you go about finding out? Who needs to be consulted when working with cultural objects? After you update and digitise your catalogue, can you make it universally accessible? Are there permissions and restrictions you should consider regarding the human remains and the religious items? Can you distinguish between collection items warranting more sensitive treatment from those that can be managed without special procedures?

Questions

The following questions are meant to be a stimulus, not an exhaustive list. They cover all sections of the book without repeating them in the same order. We organised them in such a way as you may face them in reality: start with your vision, obtain a clear understanding of the situation and consider strategic planning.

Please note that some of them might not apply to your specific situation. On the other hand, you will probably have to answer questions that are not listed below. Maybe they are already coming up while developing your strategic framework. We have left some space to add them at the end of the list.

A) Getting an overview

Your role, vision and objectives

Before you start answering the questions, consider your role as it is currently or put yourself in the position of one of the three scenarios given above.

- What is your role? What do you want or need to accomplish?
- How do your vision and objectives relate to the collection you are responsible for?
- In what ways will your work improve the collection?
- What do you want the state of the collection or museum to be in 10, 20 or 30 years' time?
- Why do you think the implementation of your vision, regular work or project is important?
- What do you consider your biggest challenges?

How will you formulate, organise and plan your objectives following your own situation or one of the three scenarios above?

- What competencies do you need in your (project) team?
- Which members of staff need to be involved?
- What should you say to members of staff who will not be involved in this project?
- Who will fill what roles and what will they contribute?
- Who is essential, who just "needs to know"?
- What if a team member leaves – do you have a backup?
- Are there any specific skills you need in your team?
- If you don't have the appropriate knowledge or skills, how will you fill the gap?
- Do you have these competencies within the organisation?
- Are there opportunities for staff development?
- Are volunteers or "the crowd" an option to include in your team?
- How will you ensure staff buy-in, if necessary?
- Which stakeholders need to be involved?

Understanding the strategic framework

What is the strategic framework and structure of your own organisation or of the organisation from the scenario you have chosen (for you to create)?

- What are the structure and decision-taking processes of the organisation?
- Do you know the mission and core objectives of the organisation?
- Can you formulate the mission and core objectives?
- Is the strategic setting still up to date? Does it support your plans and visions?

How does your strategic plan relate to the purpose, values and policies of the organization?

- How is your work or project embedded in the museum's structure?
- How does your work or the project relate to the mission?
- How will your project support the mission of the organisation?
- Does your team know the mission and core objectives?
- Are they working within the framework of the mission?
- Does your project support work towards the core objectives?
- How will your project contribute to the core objectives of the organisation?

What are your role and responsibilities in the organisation or project?

- What is your role and position in the organisation?
- What are your responsibilities? What is your authority?
- What collections are you responsible for?
- Is your role and position clear for colleagues, external users and stakeholders?

B) Your collection

Status, assessment and analyses

How well do you know the collections you are responsible for?

- Do you have an overview of the collections in your organisation?
- What are your rights and duties with respect to the collections?
- Do you know its general history and origin?
- What information is known and available about the collections you are responsible for?
- Is this information readily available and are you familiar with it?
- Is your team familiar with the history and origin of the collections?
- What is the impact your work or project will have on the collections?
- Is your team aware of this?
- Is there any relevant collection information missing that is necessary for your work or project planning and priority setting?
- How can you get that information to help you with strategic planning?

What is the status of the collections?

- Do you have methods and criteria available to analyse the collection information?
- How detailed does your assessment need to be, e.g. at specimen, storage unit or department level?
- Are there any schemas available or do you have to develop your own method?
- Are you familiar with high-level assessment methods (e.g. by SYNTHESYS) and how to use them?
- What are the possible criteria to evaluate a status of a collection?
- How can you measure them?
- How would you decide whether it is a strength or weakness?
- How would you conduct an assessment?
- What kind of data and information will the assessment provide? Do they fit to your needs? Will they answer your question?

What is the status of policies and procedures?

- Are you aware of the overall state of conservation of the collections you are responsible for?
- What are major risks to the collections when conducting your work or project, both physical (e.g. damage) and regarding its success (e.g. sustainability)?
- Are there any legal or ethical restrictions that will influence your work, e.g. health & safety issues, acquisition and deaccession or legal issues?

- Is there anything that you cannot do because of specific restrictions?
- Do you and your team have policies and procedures in place to deal with them?

How will you use the assessment data?

- How will you use the data to set your work priorities?
- What areas do the data suggest require improvement and why?
- What are your biggest challenges and why?
- What will be the follow-up to your assessment?
- How will you use the results internally and externally?
- How future-proof is your collection, i.e. how flexible are its preservation methods and accessibility to allow for future uses?

Users, stakeholders and value of collections

In order to produce relevant output, it is important in any project to understand who the stakeholders and users are and what they need. Your work should be driven by stakeholder and user needs, as our collections are. The use of collections will add value to them. Who are the major stakeholder and user groups of your collections or in the scenario you have chosen? How can you describe and increase the value of your collections?

What is the purpose of the collections?

- Why does the organisation keep these collections?
- Who are the (key) users of the collections and what do they use them for?
- What do they expect from the organisation?
- Does the organisation and collections meet the needs of these users?
- How can you measure and demonstrate (scientific, societal) impact of use of the collections?
- What are the possible criteria for measuring use?

Who are stakeholders?

- Who are the stakeholders of the organisation and/or who are stakeholders of your project?
- Who might be future stakeholders? Why and what do you have to do to link them to the organisation?

What is the value of the collections?

- How can you measure the value of collections and what are the possible criteria?

- How can you demonstrate their (scientific, societal, historic, monetary) value?
- Does the value differ for the various stakeholders?

Access to collections

- What are the benefits of access to collections?
- How can you measure it?
- How can you describe the impact of collection use on science and society?
- Can you measure the impact? If so, what measures would you use?

How is access to the collections organised and who is responsible?

- What kind of access is provided to whom and under what conditions?
- Who is taking decisions on what kind of access will be allowed and for whom?
- What are the restrictions? Why?
- What are the costs of access?
- Can you reduce the costs and if so, how?
- Are bench fees or services on demand an option and if so, how would you organise that?
- What are the risks associated with the access you provide?
- Are there ethical or cultural considerations?

Does the access to collections meet the requirements of known and potential new users?

- Who are the users and what kind of access do they need?
- Does the access you provide meet these needs?
- What are major barriers to access and how will you overcome them?
- Can you solve these barriers by small adjustments or do you need major changes?
- Is there a bottleneck in physical or digital access?
- Is digitisation a possible way to enable more customised access?
- Do you have a strategy to open up your collection digitally?
- What kind of access will users require in the future?
- How will you be prepared for this?
- How can management of the collections benefit from future or alternative means of access?

Collection development

Does your collection fit the institution's mission and/or the aim of your project?

- Do you want to optimise your collection in terms of its coverage, information content, use of space, etc.?
- What are possible ways within the legal framework?
- Will the collection you are responsible for remain static or will you acquire or dispose?
- Does your institution need to continue collecting new specimens? Why or why not?
- What are your criteria for acquisition and disposal?
- What legal requirements are there? Are there rules or laws governing this?
- Is there an existing policy in place governing what should be acquired or disposed of?
- What are the ethical considerations in acquiring or disposing of collections?
- Who should be involved in the decisions internally and/or externally?
- Do you and/or your colleagues have the levels of competency needed to make these decisions?
- Who is authorised to make these decisions?
- How will you explain to and convince the public that deaccessioning and collecting are normal and legal procedures in collections management?
- What are your alternatives to disposal?
- Is repatriation an issue and if so how are you addressing it?
- Are you acquiring objects which are taking up valuable storage space and do not really match the collections profile or benefit users?
- Do you have the resources such as staff, competencies and space to house and manage new acquisitions?
- Are there any health and safety considerations in play?

C) *Project planning*

Human resources

How will you organise your work and the team you are about to lead?

- How will you organise the decision-taking processes?
- Are roles and responsibilities clear?
- Do your team members know about their roles and responsibilities?
- Does the structure allow for communication, knowledge sharing and help you to achieve your goals?
- If not, can you change it?
- If you cannot change it, what mechanisms can you put into place to help you achieve your goals?

What personnel resources do you need to achieve your goals?

- What specialist skills and competencies do you need in your team?
- Do you have these skills and competencies in your team?

- What kind of competencies are missing but essential to add to your team?
- How will you build a team with the competencies you need?
- Is there a pool of resources you can use?
- Are there staff members that can be developed in a way needed for the project?
- Will you need to bring in or ask for new staff?
- How will you fund that?

Are there other ways to acquire necessary competencies and expertise?

- Do you need to bring in external expertise? How would you do this?
- Is outsourcing an option?
- Are volunteers an option? How would you organise this?
- Could your objectives be met through collaboration with other institutions?
- Are there existing consortia? How would you approach them?
- Are there specialist groups that could help with advice? How would you engage with them?

Budget and fund raising

What financial resources do you need to realise your project?

- Is all information available to make proper calculations?
- What funds are available?
- Can you get sufficient funds to realise your plans?
- If not, what are your priorities, what would you cancel or postpone and why?
- How is your decision-making process organised, if you will have to scale down?
- Are there any opportunities to be more cost efficient (e.g. outsourcing versus in-house, off-site versus on-site storage or collaboration and consortia) to achieve your goals?

Do you have a strategy to find additional funds to realise your project?

- Do you know where to look for additional funding?
- Do you have an overview of funding agencies and funding programmes on local, national and international levels?
- Do you have a team member with fund raising skills?
- What will be your fund raising strategy?
- How will you sell your project to potential funders or to your Director?
- Do you have a proper project proposal?
- Are you prepared to pitch your idea to a potential funder at any time?

Performance indicators, risks and controlling

How will you measure progress and success of your project/activities/collection performance?

- What indicators do you currently use?
- Do they measure performance against objectives?
- Are they measuring activity or performance?
- Are they meaningful and not conflicting with other indicators?
- What other indicators do you need?

What are the risks to the success of your project?

- Have you conducted a risk assessment?
- Do you have a risk log?
- Are there any legal or ethical constraints that might affect your project, e.g. health and safety?
- What barriers or risks have been identified, e.g. is the infrastructure such as IT services in place to support your objectives?
- How might the risks influence or interfere with your project/goals/objectives?
- What kind of preventive measures should you take?
- What would be the worst case scenario?
- Are you prepared for this and what would you do?
- What kind of controlling mechanisms and safety measures do you have in place to prevent the project from failing?

D) *Your situation and your questions*

When developing your own strategic framework, you will face more specific questions. There is space here to describe your situation and add your questions. This could be helpful when practicing, but also as a reminder when looking back at a later stage in your career.

Your situation and your challenges

What is your challenge and/or project? What are the important aspects that you need to keep in mind?

- …
- …
- …
- …

Which questions are missing?

What are the specific questions that you had to or will have to ask (and have answered) in your situation?

- …
- …
- …
- …

A few final words

We hope that by sharing our experiences and stepping through the strategic management process with you we have helped you in your current position or supported you in making the next step in your career. You may now want to think about what to do next after you have read the book and worked through the questions.

We learned a lot ourselves through participating in, and later leading on, reviews and assessments of a wide variety of collections around the world – a great way to learn not only how to do things but also how not to do them! But also, and perhaps most importantly, we have been privileged to be part of many networks and societies which gave us the opportunity to share ideas on the challenges of our collections with colleagues and friends. We would urge you to engage with your colleagues within and outside of your institution and also with those who may manage very different collections to your own and can bring new perspectives to your ideas and issues.

Good luck and we hope that you will enjoy the challenges and excitement as much as we did.

Glossary

For a comprehensive list of terms and alternative definitions in collections management and preservation see *Museum Wise: Workplace Words Defined* (Cato, Golden & McLaren, 2003).

Access: studying specimens directly in person by visits to collections, by borrowing material (loans) or virtually through examination of digital surrogates.

Access and benefit sharing (ABS)/Nagoya Protocol: the Nagoya Protocol on Access to Genetic Resources and the Fair and Equitable Sharing of Benefits Arising from their Utilization of the Convention on Biological Diversity (in short, access and benefit sharing, ABS) aims at regulating the access to genetic resources and the fair and equitable sharing of benefits derived from their utilisation. The fair and equitable sharing of the benefits arising out of the utilisation of genetic resources is one of the three objectives of the Convention on Biological Diversity (CBD). See www.diversitas-international.org/activities/policy/cbd-1/access-and-benefits-sharing-abs.

Accessioning: the formal process of legally accepting and recording an object into an institution's collection. The final process of acquisition. Often used interchangeably with ***Acquisition***.

Acquisition: taking legal ownership of objects, especially to add to your long-term collection through the process of accessioning. The formal commitment by your governing body to care for objects over the long term (see Spectrum: https://collectionstrust.org.uk/spectrum/). Often used interchangeably with ***Accessioning***.

Agents of deterioration: risks facing museum collections: fire, pests, direct physical forces, water, incorrect temperature, incorrect relative humidity, ultraviolet and infrared light, pollutants, dissociation and thieves and vandals.

Annotations: additional information added to a specimen, such as name and type status, usually with the name of the annotator.

Behaviours: in the context of this book, an individual's personal characteristics relevant to the job such as ability to work with visitors, work in teams, etc.

Bench fees: fees to cover the use of the organisation's facilities and services, such as an area to work, staff assistance, locating relevant literature, tools such as microscopes, internet access, etc. Some organisations either partially or completely waive these in return for improvement to the host's collections, such as specimen determinations, typification, updating taxonomy, etc.

Best practice: the process of developing and following a standard way of doing things in the best way possible and in a way that multiple organisations can use.

Big Data: very large sets of specimen data which, when analysed, can reveal patterns such as changes in distributions of organisms over time.

Board of trustees: an appointed or elected committee that oversees the business of a museum.

CETAF: see ***Consortium of European Taxonomic Facilities (CETAF)***.

Collection management system: software that manages all aspects of a collection and its items from accessioning and deaccessioning through to incoming and outgoing loans, shipments, location tracking and internal movements, valuations, insurance and indemnity, and more.

Collection manager: person responsible for the day to day management of a collection. In some institutions this role is referred to as "curator" and the term used for a more senior position managing a team of "curators" (e.g. Natural History Museum, London).

Collection profile: the composition of a collection based on its history and past, present and future research objectives. Collection profiles need to be formulated in a collection management plan which is an integral part of an institutional management plan.

Competency framework: a structured way of recording the competencies required by staff to fulfil their roles in an organisation.

Conservation: actions taken to maximise an object's useful life or minimise deterioration.

Conservator: a person trained in the preventative care and restoration of museum objects.

Consortium of European Taxonomic Facilities (CETAF): a networked consortium of scientific institutions in Europe formed to promote training, research and understanding of systematic biology and palaeobiology. Together, CETAF institutions hold very substantial biological (zoological and botanical), palaeobiological and geological collections and provide the resource for the work of thousands of researchers in a variety of scientific disciplines (www.cetaf.org).

Convention on Biological Diversity (CBD): a global agreement addressing all aspects of biological diversity: genetic resources, species and ecosystems. The objectives of this convention, to be pursued in accordance with its relevant provisions, are 1) the conservation of biological diversity, 2) the sustainable use of its components and 3) the fair and equitable sharing of the benefits arising out of the utilisation of genetic resources, including by appropriate access to genetic resources and by appropriate transfer of relevant technologies, taking into account all rights over those resources and to technologies, and by appropriate funding (www.cbd.int/).

Core competency: a fundamental knowledge, ability, or expertise in a specific subject area or skill set. In the competency framework, they represent a set of essential competencies of an individual required to meet the expectations of a role.

Crowd-sourcing: gathering and recording information by using the input of the general public, usually on-line, for example transcribing data from images of labels.

Curator: a term variably applied. In this book, a senior researcher usually in charge of a department or a person responsible for the management of a specific collection, e.g. curator of ferns (see also ***Collection manager)***

Deaccession/Deaccessioning: the process of approving and recording the removal of objects from a museum collection.

Digitising: the process of creating a database and/or digital versions (images, 3D models or audio-visual files) of objects in collections and making them available for use.

Disaster planning: usually a written plan that specifies procedures to follow in the event of an emergency such as an incident that threatens human safety and damages or threatens to damage facilities and resources such as the collection.

Disposal: the act of physically removing an object or collection from its institution/owner. Types of disposal include donation, transfer, exchange, sale, repatriation, sampling for destructive analysis, consumptive or educational use and destruction.

EDIT (European Distributed Institute of Taxonomy): a European Commission-funded "Network of Excellence" project aimed at reducing fragmentation in European taxonomy through increased cooperation and integration between the larger taxonomic research institutions. EDIT explored new ways of institutional collaboration and, in particular, the use of innovative working methods and the integration of taxonomic data and information through the internet (cybertaxonomy).

FAIR: Findable, Accessible, Interoperable and Re-usable. See www.force11.org/group/fairgroup/fairprinciples.

GBIF: see ***Global Biodiversity Information Facility (GBIF)***.

General Data Protection Regulation (GDPR): a regulation in EU law on data protection and privacy for all individuals within the European

Union (EU) and the European Economic Area (EEA). It also addresses the export of personal data outside the EU and EEA areas.

Georeferencing: assigning geographic coordinates to a location. For example converting a place name on a specimen label to coordinates.

Global Biodiversity Information Facility (GBIF): an international open data infrastructure, funded by governments. It allows open access to data about all types of life on Earth (e.g. distribution data on plants, animals, fungi and microbes of the world, and scientific names data), shared via the internet, by encouraging and helping institutions to publish data according to common standards (www.gbif.org).

Governance: the array of laws, regulations, policies, practices, rules and procedures that guide the work of the museum.

ICOM: International Council of Museums (https://icom.museum/).

Intellectual framework: a description of the nature of a collection, what it contains and why it exists.

Key Performance Indicator (KPI): a type of performance measurement and a way for an organisation to evaluate the success of a particular activity or progress toward strategic goals. For example numbers of loans, visitors in a year, number of specimens conserved/digitised, etc.

KPI: see ***Key Performance Indicator.***

Mission statement: a short statement of why an organisation, such as a museum, exists, its overall goal and the services it provides.

Mutually Agreed Terms (MAT): in the context of ABS and the Nagoya Protocol, a contractual arrangement establishing that specific benefit-sharing conditions must be "mutually agreed" between providers and users of genetic resources (includes most natural science collections as they can carry genetic material potentially useful to humans).

Nagoya Protocol: see under ***Access and benefit sharing.***

Object: collections item. Objects in natural science collections are commonly referred to as "specimens".

Optical Character Recognition (OCR): computer technology that recognises printed or handwritten text in digital images, such as a scanned specimen label.

Open access policy: a mechanism allowing access to data to all free of cost or other barriers.

Orphaned collections: collections that no longer have curatorial support or have been abandoned by their owner.

Pattern recognition: automated recognition of patterns and regularities in data.

Performance indicator: see ***Key Performance Indicator.***

Preservation: actions taken to retard or prevent deterioration or damage to collections materials by control of their environment and/or treatment of their structure. The non-invasive act of minimising deterioration and preventing future damage to an object.

Prior Informed Consent (PIC): in the context of ABS and the Nagoya Protocol, PIC refers to the administrative permit given by the competent national authority of a provider country to a user, prior to accessing genetic resources. The term can also be used in relation to the right of indigenous and local communities to give access to genetic resources or associated traditional knowledge.

Public engagement activities: in a museum's context this usually refers to public exhibitions and educational activities.

Registrar: in collections management, the person/office that deals with compliance with rules and regulations governing specimen transactions, negotiation of terms for loans, etc. Most art/humanities and general museums have such an office, but it is less common in natural history museums.

Re-spiriting: the act of adding more fluid preservative to a container. Can also be called "topping off" or "topping up".

SPNHC: Society for the Preservation of Natural History Collections (www.spnhc.org/).

Stakeholder: someone who has an interest in a topic, problem or situation and a role to play in developing the response or solution, such as users of the collections, funding and governing bodies.

SYNTHESYS: Synthesis of Systematic Resources, a series of European Commission-funded projects, creating an integrated European infrastructure for natural history collections (www.synthesys.info/).

Taxonomy: a branch of systematics, namely the description, identification, naming and classification of organisms.

Taxonomist: a professional or amateur scientist describing, identifying, naming and arranging organisms.

Taxonomic Databases Working Group (TDWG): a not-for-profit scientific and educational association that is affiliated with the International Union of Biological Sciences (www.tdwg.org).

Taxonomic group: a unit of biological classification above species level, used for organisational purposes in a natural science collection, such as family or order.

TDWG: see ***Taxonomic Databases Working Group (TDWG).***

Time series: representations of specimens sampled over time, either by conscious collecting strategy or by chance.

Type/Type specimen: the specimen or a set of specimens on which the description and name of a new species is based and formally attached.

Users: a subset of stakeholders who study or use our collections, for example scientific visitors, borrowers and exhibitions staff.

Value chain: a set of activities that an organisation such as a museum performs in order to deliver a valuable product to its stakeholders.

Visitor: user who visits the collection to study it.

Voucher specimen: a preserved specimen that serves as the supporting material for published studies on that species and ensures that the science is repeatable.

Reference

Cato, P., Golden, J., and McLaren, S.B., 2003, *MuseumWise* : *Workplace Words Defined*. Washington, DC: Society for the Preservation of Natural History Collections.

Index

Note: Page numbers in *italics* refer to illustrations and those in **bold** refer to tables. Footnotes are denoted by n, e.g. 206n6.